PENGUIN BOOKS

A TEASPOON AND AN OPEN MIND

Michael White is the author of some twenty-five books, which have appeared in over 150 editions around the world. His titles include the international bestsellers *Stephen Hawking: A Life in Science*, *Leonardo: The First Scientist*, *Tolkien: A Biography* and *The Science of the X-Files*. He divides his time between London and Perth, Western Australia.

For more information visit michaelwhite.com.au

GW00507296

MICHAEL WHITE

A Teaspoon and an Open Mind

What would an alien look like?
Is time travel possible?
and other intergalactic conundrums from
the world of Doctor Who

PENGUIN BOOKS

PENGUIN BOOKS

Published by the Penguin Group
Penguin Books Ltd, 80 Strand, London WC2R 0RL, England
Penguin Group (USA) Inc., 375 Hudson Street, New York, New York 10014, USA
Penguin Group (Canada), 90 Eglinton Avenue East, Suite 700, Toronto, Ontario, Canada M4P 2Y3
(a division of Pearson Penguin Canada Inc.)
Penguin Ireland, 25 St Stephen's Green, Dublin 2, Ireland
(a division of Penguin Books Ltd)
Penguin Group (Australia), 250 Camberwell Road, Camberwell, Victoria 3124, Australia
(a division of Pearson Australia Group Pty Ltd)
Penguin Books India Pvt Ltd, 11 Community Centre, Panchsheel Park, New Delhi – 110 017, India
Penguin Group (NZ), 67 Apollo Drive, Mairangi Bay, Auckland 1310, New Zealand
(a division of Pearson New Zealand Ltd)
Penguin Books (South Africa) (Pty) Ltd, 24 Sturdee Avenue,
Rosebank, Johannesburg 2196, South Africa

Penguin Books Ltd, Registered Offices: 80 Strand, London WC2R 0RL, England

www.penguin.com

First published by Allen Lane 2005
Published in Penguin Books 2006
1

Copyright © Michael White, 2005
All rights reserved

The moral right of the author has been asserted

Typeset by Rowland Phototypesetting Ltd, Bury St Edmunds, Suffolk
Printed in England by Clays Ltd, St Ives plc

Except in the United States of America, this book is sold subject
to the condition that it shall not, by way of trade or otherwise, be lent,
re-sold, hired out, or otherwise circulated without the publisher's
prior consent in any form of binding or cover other than that in
which it is published and without a similar condition including this
condition being imposed on the subsequent purchaser

ISBN-13: 978–0–141–02481–3
ISBN-10: 0–141–02481–X

Contents

For Michael Downer and Nicolas Peglitsis
– the best teachers I ever had

Introduction

I had to make some optimistic assumptions to meet the revenue target. In Week Three, we're visited by an alien named D'utox Inag who offers to share his advanced technology.   Dilbert

In the late 1990s, my wife Lisa and I did the clichéd thing of leaving London to raise our kids in the country. We lived in an old barn just outside a town called Headcorn in Kent, and within a few months of being there we discovered that Tom Baker lived in the next village. I recall fondly the time Lisa came back from town to say, 'You'll never guess who I saw in Headcorn . . . Doctor Who!' At least once a month she would bump into the Doctor in the local butcher's shop or the florist's, where she would overhear his often hilarious banter with the shop assistants.

Every *Doctor Who* fan has a favourite Doctor, and mine were Tom Baker and Jon Pertwee. The choice of favourite is always connected with one's age. If you're a child of the seventies then you will almost certainly have the same favourites as me. But actually my interest in *Doctor Who* goes further back, to the mid 1960s, when as a very small child I was petrified by the Cybermen and the Daleks and hid behind the sofa as soon as that familiar synthesizer tune (courtesy of the BBC Radiophonic Workshop) started playing every Saturday evening.

Later, I became a fan of science fiction and read every Asimov novel, everything by Arthur C. Clarke and Robert Heinlein and others, but always retaining a soft spot for *Doctor Who*. In some ways the programme was not very good science fiction. It didn't have the depth and accuracy of the novelists I liked, it wasn't so big-budget as *Star Trek*, and at the age of ten or eleven I found it easier to relate to Will Robinson in *Lost in Space*. But *Doctor Who* always had a sense of humour, and this was brought out best by Tom Baker with his scarves and jelly babies.

I have also always loved the quirkiness and the very British aspects of the programme. It is as British as cups of tea, the Kinks and losing in the final stages of the World Cup. I'm sure this is part of its enduring appeal. The Doctor has always been cool whether he wears frilly shirts and drives around in a yellow roadster or sports a leather jacket and trainers. The way he deals with problems, Heath Robinson style, blending the exotic with the mundane and clutching victory from the jaws of defeat, is one of his most attractive traits.

Science fiction is a much-maligned form, and literary purists turn their noses up at it (while at the same time shamelessly nicking ideas from the genre). But science fiction has a place of enormous importance in the lives of many people, most especially all those brave souls who become scientists. No less a figure than Stephen Hawking has said that he was originally drawn to study science because of his interest in science fiction. Many of the great science-fiction authors of the twentieth century did the same and then went on to become writers themselves who in turn led generations of young people to science.

It is obvious why this should happen. Science fiction is really 'super-science', science taken further than its present state, imaginative extrapolations into an unknown future. Science fiction is liberating. It acts as a vehicle for escapism, of course, but more than this, it opens up a bigger vista, a more exciting reality than our often humdrum lives can offer. By moving into real science we try to recapture the excitement and the glamour of what we have read in novels.

The great thing about science is that it is always changing and growing. Along with the fact that science is based on experiment and analysis, it is this ability to develop that really separates it from mere belief systems. But, even though science evolves, every good scientist goes with the flow.

An example of this is the changing view of the nature of human genetics. Until very recently it was believed that the only useful part of the genome were the genes themselves. But scientists have now found that what was originally called 'junk DNA' – usually viewed as merely the padding between genes – may be vitally important. Indeed these regions of the genome have been preserved in every species studied and have not changed over millions of years, implying that they are somehow protected or preserved for a special purpose.

This is almost certainly an exciting and important discovery, but it is also quite possible that in a year or two or five another scientist or team of researchers will develop a subtle refinement of this work or even perhaps discover something that refutes it. This is the way of science and one of the things that keeps it vital and alive. It does not mean scientists have lost their grip on reality or that science is in any way subjective. It simply means that scientific thinking develops. As Asimov once said: 'When people thought the Earth was flat, they were wrong. When people thought the Earth was spherical they were wrong. But if you think that thinking the Earth is spherical is just as wrong as thinking the Earth is flat, then your view is wronger than both of them put together.'

So, what am I trying to do with the book you have in your hands? It is a book that owes much to *Doctor Who*, but it is not really about the programme or the characters involved in the stories. Instead, it is about the science behind the stories. In it I take a serious look at what might be possible and what is most likely impossible, a look at the potential of science to fulfil the dreams of science-fiction fans.

For anyone interested in considering the outer limits of science, it is, as the title of this book suggests, important to keep an open mind. One could list hundreds of quotes offering misguided scientific judgements about the future. My personal favourite is from Lord Kelvin, who declared, 'Heavier-than-air flying machines are impossible.' He made this remark in 1892 (just eleven years before the Wright brothers' first flight), and at the time he was the most respected scientist in the world.

One of the lessons to be learned from this is the central credo of this book – that with science we should always expect the unexpected.

Science is not just constantly evolving: it throws us curves and keeps us on our toes. An appreciation of science, as I hope this book will prove, helps us to believe that almost anything is possible.

Michael White
Perth, Australia
August 2005

1

Secrets of the Time Lords
Is Time Travel Possible?

Time is a sort of river of passing events, and strong is its current; no sooner is a thing brought to sight than it is swept by and another takes its place, and this too will be swept away.   Roman emperor Marcus Aurelius

The core idea of *Doctor Who* is that our hero, the Doctor, is a Time Lord, an advanced alien being who is able to travel through time, into the past and into the future. It is a concept that has fascinated science-fiction and fantasy writers and their readership for generations. Time travel is a central theme in H. G. Wells' *The Time Machine*, the TV series *Stargate* and movies such as the *Back to the Future* series, *Twelve Monkeys* and *Time and Again*.

Most of us have wondered where we would go and what we would do if we had the ability to return to the past or to travel into the future. Would you want to know what is going to happen to you – who you will marry, what your kids will be like? Would you wish to return to the past and correct a mistake, unsay something you should never have said? Or alternatively, would you want to experience for yourself the Battle of Waterloo or the Kennedy assassination? Some people would instead like to see what will befall the human race, to witness what wonders and horrors lie ahead of us.

But is any of this really possible? Is time travel an idea that will always remain fantasy or is there a chance that in the distant future we might develop the technology to travel in time? Alternatively, is it feasible that there exists an advanced civilization somewhere in the universe that is even now able to master time and to travel anywhen at will?

At the moment, no one has a clear idea how a time machine could be constructed, and physicists working at the very edge of science are only now beginning to piece together theories that may explain how time travel could be possible in some distant future. Today, these ideas exist only as mathematical concepts. In practical terms we are a very long way from building a Tardis. But, the initial step towards constructing a time machine is to understand the mathematics behind it. And before we can develop any theory of time travel we have to get to grips with the meaning of time itself.

We all experience the passing of time, but no one seems able to explain conclusively what time is. Some even suggest it is nothing more than a construct of our own minds, that we piece together events in a logical, linear order because that is the only way our brains can operate and make sense of the universe. There is no hard evidence to support this concept but, for what it's worth, we all seem to have an in-built awareness of the direction of time, a concept which has been dubbed the *arrow of time*. But, beyond this subjective perception of time, the more clinical answers provided by physics may eventually lead to a time machine.

It is a striking fact that on a sub-atomic level almost all processes in the universe can be conducted in either temporal direction. This means that if two sub-atomic particles come together and interact to form two other particles, the reverse of this process is equally viable: the two product particles could just as well interact to create the starting particles.

Yet we don't experience this reversibility in the 'real' world of our everyday existence. We don't see shattered glasses re-form, light does not leave our eyes and travel to distant objects, and the dead do not rise from their graves. Yet this seems like a contradiction, because it implies that principles governing the behaviour of 'simple' systems (those that operate on a quantum level, such as atoms) are not the same as those that determine things in our everyday lives – such as how a car moves or the motion of a billiard ball. This is weird and seems paradoxical because, after all, every material thing in the

2

universe is made up of fundamental particles. If these particles behave reversibly on the simple atomic scale, what is it about everyday situations that seems to make them act differently?

The answer lies in the difference between something being *impossible* and just very *unlikely*. Physicists believe that it is not impossible that the dead could be made to rise again (ignoring spiritual considerations), or for a broken glass to re-form by chance; it is just that these events require so many improbable steps to interlink perfectly (at least compared to the interaction of two sub-atomic particles) that we would almost certainly have to wait for a period longer than the lifetime of the universe to see them happen naturally. This means that although they are not *impossible*, they are highly *improbable*.

Another way of putting this is to say that quite literally anything can happen, but the more unlikely it is, the longer you would have to wait to witness it. It is possible you could spontaneously transform into a jelly fish while reading this book, but it is unlikely. However, if you had many times the life span of our universe in which to read the book, there is a more significant chance that during that period you would transform into a jelly fish.

To see why this should be, we need to consider one of the most fundamental rules of the universe, a principle called the second law of thermodynamics.

This law lies at the very heart of physics, and, unlike many ideas in science, the second law of thermodynamics is actually one based entirely upon common sense. Put simply it says that 'Everything wears out', or, in more formal terms, 'The entropy of a *closed system* always increases'.

The word 'entropy' is the technical term describing the 'level of disorder in a system'. So, by the second law of thermodynamics, a cup of tea exhibits a higher level of entropy than the individual tea leaves, water and milk because they have been mixed together – the mixture is more *disordered* than the original ingredients in their separate containers. Another way of thinking about it is that it would take more energy to separate out the ingredients of the tea and put them back in their individual containers than it took to mix them up together in the first place.

In the case of the broken glass I mentioned earlier, if we tried to bring together the pieces, like running a film backwards to re-create the glass perfectly, we would need to lower the entropy of the system. This is possible (in fact, living creatures spend most of their time attempting to produce a local lowering of entropy – just think of housework as an example), but it requires energy (in this case, more energy than it took to shatter the glass), and so the chance of this happening naturally is incredibly small.

Consider another example. A garden left to overgrow will gradually increase its entropy level quite naturally. In order to restore the tangled weeds and vines to their former order (to lower the entropy) work would have to be done or energy expended (again, more work or energy would be needed than Nature used in disordering the garden). It is extremely unlikely this will happen naturally without applying energy.

Because the natural processes of the universe are all ones in which disorder or entropy is seen to be increasing, it gives us an indicator, a way to view the evolution of the universe or, in other words, the direction in which time flows. 'The arrow of time' points in the direction of increased entropy; what we call 'the future'.

So, having established this, if we assume there is a definite direction to time, can intelligent beings move in a non-linear way from one time frame to another? Would it ever be possible to build a machine like the Tardis?

Currently, physicists are giving serious consideration to a small collection of possible mechanisms via which genuine time travel may be accomplished. At present all of these ideas lie way beyond the bounds of early 21st century technology and are based on some pretty far-out physics as well as a sprinkling of informed speculation, but they are at least a start, an attempt to get to grips with the idea.

The first and, for the moment, the best idea is for a time traveller to use the services of a thing called a *wormhole*, an idea which arose as a consequence of manipulating the mathematics of Einstein's theory of relativity.

There are actually two parts to the theory of relativity, the *special theory* of 1905, written when Einstein was working in a Bern patent office, and the *general theory*, first revealed to the world eleven years later in 1916, which is an extension of the more limited special relativity.

The special theory draws upon two firmly established scientific principles, but it also comes up with another which many non-scientists consider to be one of the weirdest notions in the whole of science.

The first principle employed by Einstein comes from the work of Isaac Newton, who, back in the 1680s, showed that the laws of physics are the same for any observers moving at a constant velocity relative to one another. So if a driver and passengers in a car travel alongside another car (or travel towards the other car), and both cars are travelling at constant velocity, the driver and passengers in each vehicle will each perceive the universe to be behaving in the same way. This seems obvious, but it has important ramifications for Einstein's concepts.

The second fact is that the speed of light in a vacuum is always constant. This velocity is represented by the symbol c and is equal to about 300,000 kilometres per second, or just over one billion kilometres per hour. But, most importantly, this value is the same *irrespective* of the velocity of the observer.

According to common sense, if, say, spaceship A is moving in one direction with a velocity of 0.75 c and spaceship B approaches in the opposite direction also travelling at 0.75 c, their relative velocity would be 1.5 c. But this is not actually the case. According to Einstein's equations, crews on each ship would see light from the other coming towards them, not at one and a half times the speed of light, but at just under 1 c (0.96 c to be precise).

The astonishing consequence of this is that if c is constant, space and time must be relative. In other words, if the crew aboard spaceship A or B are to see light arriving at a constant velocity irrespective of their own velocity, they must measure time differently – as they travel faster, time must slow down. Furthermore, because distance, time and speed are all interrelated, if time slows down, then the property of distance cannot be the same for observers travelling at different speeds. The faster one travels, the shorter any given spatial measurement becomes – a metre will be a different length depending on the velocity of the observer, and it will be shorter the faster the observer moves.

Finally, the faster an observer moves, the more massive they become. The end result of all this is that if it were possible for an observer to travel at the speed of light they would experience three rather strange things simultaneously – to them time would slow to nothing, they would shrink to nothing and their mass would be infinite. And this is no mad theory or unproven hypothesis. Einstein's work has been shown to be true in many thousands of experiments conducted since 1905.

This, then, is the special theory of relativity; it is only concerned with observers moving at a constant velocity, but after it was published Einstein wanted to know how the concept of relativity might work for objects experiencing acceleration.

He worked through a thought experiment – that is, he imagined a situation rather than actually performing an experiment. He needed to visualize a situation in which an object was accelerating rather than just moving at a constant velocity. To do this he imagined a lift in a state of free fall. Thanks to gravity, this lift would be accelerating towards the ground. Next, he imagined a beam of light fired at the side of the lift and entering a hole in one wall.

Now, imagine you are in the lift. It would seem to you that the light beam comes in through the hole and strikes the opposite wall at the same height. In other words, it would look as though the beam of light was straight.

However, if you were outside the lift and watching this process, the light would appear bent. This is because in the time it took for the light to cross the width of the lift, the lift has moved downward in free fall, so the light would be seen to hit the far wall at a point higher than the hole in which the light entered. So, to you, the external observer, the light beam would appear bent.

But, said Einstein, there is a problem with this. As with special relativity, the behaviour of light must be the same for all observers unless proven otherwise. This idea is called the 'principle of equivalence', and it is as sacrosanct as any scientific idea can be, so that other factors have to be altered to accommodate it. In order to uphold the principle the only logical conclusion is that space itself is warped by the force of gravity (which is providing the acceleration of the lift) and this causes the bending of light. The stronger the gravitational effect, the more the light would be bent. So, if the lift was falling to the surface of a planet where the gravitational field was much stronger than that on Earth, the light would appear to be more bent than in the original thought experiment.

This was the essence of Einstein's 1916 paper describing the general theory of relativity, and only three years later it was proved correct using a 'real experiment' conducted by the great British astronomer Arthur Eddington. During an eclipse, Eddington observed how light from a distant star was bent by the gravitational effect of our sun. This happens because the sun distorts the fabric of the universe around it, which then bends the path of observed light.

Until Einstein, physicists saw the universe in three dimensions, with time as an extra factor. In general relativity, time is a dimension just like length, breadth or depth; the universe actually exists in four dimensions called 'space-time'.

The only way we can visualize a four-dimensional universe is by representing it in three dimensions. Imagine a rubber sheet stretched flat. Now place a heavy ball in the middle – the sheet around the ball is misshapen the way space-time distorts around a massive object like our sun. Roll a marble along the sheet near the heavy ball and it follows a curved path, just as light does near our sun or any star.

A *black hole* is so massive and has such a powerful gravitational field that it curves space so much that within it lies what is called a 'singularity', a point at which the curvature of space-time becomes infinitely sharp and all the laws of physics break down. *Wormholes*, as theorized by a number of scientists, are created when two singularities 'find' each other and join up. But do such things as black holes and singularities really exist? And if so, how could they be formed?

Scientists have known for a long time that when a star has used almost all its available fuel it begins to die, and the way in which it dies depends upon its mass. If it is about three times the mass of our sun or larger it begins to shrink, setting up shock waves which result in an enormous explosion – a supernova. But even then, because the star was so large to begin with, some material is left at the centre of the supernova and this begins to collapse in upon itself again. This time, the matter becomes so dense that the incredibly strong forces holding sub-atomic particles together, the binding forces between quarks (the most fundamental form of matter known and the constituents of protons, neutrons and electrons), are overwhelmed and the star becomes a seething cauldron of fundamental matter and energy. This is a black hole, so called because it is so massive and dense that even light (which travels at about 300,000 kilometres per second) cannot move fast enough to escape its gravitational field.

It often happens in science that mathematics predicts that something should exist and what its properties are before it is observed. Although the existence of black holes has not yet been confirmed beyond doubt, there are some promising candidates, and it is more than likely that they do exist somewhere in the universe. The chances of finding wormholes is slimmer, but there is nothing within the laws of physics that says they could not exist.

As we shall see in the next chapter, the wormhole is more commonly proposed as a way of traversing interstellar distances and to circumvent the frustrating consequence of Einstein's theory of relativity – that there is a speed limit (the speed of light) to the universe. Wormholes may be thought of as tunnels or links between two different parts of the universe through which an intrepid space traveller might pass. But wormholes might also be of great use to the potential time traveller. According to a theory first postulated by Kip Thorne at Caltech, it might be possible to create special wormholes that can provide the means to travel through time.

Thorne's effort to investigate the potential of wormholes was actually sparked by a piece of science fiction. In 1983, the late physicist, novelist and popularizer Carl Sagan was working up a concept for a novel he planned to call *Contact* (1985). In the story, an alien race sends a signal to earth containing details of how to build an interstellar craft that can travel almost instantaneously to a planet orbiting the star Vega, 26 light years from earth. Sagan was of course aware of the problem presented by the universal speed limit, and he understood that some odd and lateral method of travelling across such a vast distance would have to be contrived to make his story work. For help he turned to his friend Kip Thorne, then one of a mere handful of black hole experts and a man who relished solving such puzzles as the one Sagan was offering.

In the conversations between Sagan and Thorne, the novelist had made it clear that he wanted his science fiction to be as scientifically accurate as possible – far-fetched was fine, but the concepts he employed had to be scientifically plausible. Yet both men knew the tremendous difficulties involved in using black holes and wormholes for anything. The inside of a black hole is probably the most inhospitable place in the universe; the gravitational forces at work there would instantly break any material object into a soup of fundamental particles and energy, and, even if these forces could be resisted, once within the grip of the black hole there is no escape because nothing can overcome its gravitational field. So the idea of using a wormhole created by joining two black holes at two different points in the universe did not seem very practical. The only way they could be used would be if there happen to be certain types of black hole somewhere in the universe that do allow safe passage – an extremely unrealistic idea.

A possible way around this difficulty suggested by Thorne involves the concept of *white holes*. These would be the very opposite of black holes: rather than absorbing matter and energy, such objects might act as perfect emitters or 'cosmic gushers'. If a black hole and a white hole were joined they could act as a one-way wormhole and solve the problem of escaping a black hole once it has been entered.

Unfortunately, detailed mathematical analysis of this scenario has shown that such a system would be unstable and the white hole would rapidly decay, making the passage of a spaceship or time traveller impossible.

Thorne realized that, to construct a workable wormhole, a set of strict conditions had to be met. These included the obvious fact that the construction of the wormhole must be consistent with general relativity and that the gravitational 'tidal forces' within the wormhole be kept to a minimum. The rules also stipulated the shape to which the wormhole must conform and the mass and type of material needed to create it. Unfortunately, the mathematics showed that in order to construct a wormhole, material known as 'exotic matter', which has the bizarre property of negative mass, is needed.

For a science-fiction novel, such speculation was enough, but in thinking about the concept, Kip Thorne had been hooked and he went on to consider seriously the concept of the wormhole and its uses. In 1987, he and a colleague at Caltech, Michael Morris, published a scientific paper on the subject in the *American Journal of Physics*. This became a landmark piece of work and sparked the interest of a generation of investigators who have since studied the theoretical properties of black holes, white holes and wormholes.

But how does this relate to time travel? Well, if we treat the two ends of a wormhole as two separate objects and exploit the difference between their relative velocities, we can make one end exist in a different time from the other. Remember, time is relative. If one end of a wormhole is made to travel at close to the speed of light, and the other is kept stationary, then according to Einstein's special theory of relativity time will pass more slowly for the high-speed wormhole. In effect, relative to the stationary one, the fast-moving wormhole will be existing in the past.

But how can this help? If we are not even sure wormholes exist, what good is this concept?

First, as a consequence of studies made since the mid 1980s it seems likely that mini-wormholes do exist and that they may be found all around us. Stephen Hawking and others have speculated that mini-wormholes could provide links between different places and times. According to this theory at least, it might well be that wormholes are more common than we think. It is even possible that our universe is laced with a vast network of interlinking wormholes just waiting to be exploited by a sufficiently advanced civilization.

Another possibility is the idea that artificial man-made wormholes could be constructed. If we consider the sort of forces employed by Nature to create such entities it makes the very notion of manufacturing a wormhole seem rather ridiculous, but physicists insist that (in theory at least) such a thing is possible. It is simply that at present the technology required to produce such a wonder is almost beyond imagining.

If we want to make a wormhole we must employ very large amounts of energy. This is needed to manipulate the sub-atomic particles so that we can use Einstein's special relativity to create a time difference between the two portals. One estimate suggests that the energy required to manufacture a wormhole of a practical size for use by a time traveller would be greater than that consumed by humankind during the past two thousand years.

Even so, say the enthusiasts, if this difficulty could be surmounted and artificial wormholes could be created, they might be employed as time machines. To do this the experimenter needs to place one end of the wormhole in a cyclotron. There it would be accelerated to close to the speed of light and left to spin for, say, five years. Because time would be passing far slower for the high-speed end of the wormhole, after five years a traveller leaving in 2010 could make their way through the link and exit the other end almost five years earlier (it wouldn't be exactly five years as the wormhole could only be accelerated to close to the speed of light).

Although this sounds rather crazy, there are some sound suggestions for ways in which a wormhole may be constructed for use in such an experiment. But first, to understand these proposals we must consider yet another fundamental theory of physics, quantum theory: like Einstein's theories of relativity, quantum theory may be viewed as a supporting pillar of modern science.

Quantum mechanics began life during the early part of the 20th century and came as a completely revolutionary theory, overturning the prevailing (classical) ideas of Victorian physicists.

The earliest model for the atomic world held that the atom was composed of a nucleus around which electrons orbited – like a solar system in miniature. Electrons were known to have about 1/2000th the mass of a proton (one of the constituents of the nucleus) and to possess a negative charge to counter the positive charge of the proton.

But during the first decades of the 20th century it was realized that this model could not possibly work. For a start, mathematics demonstrated that electrons could not be sustained in their orbits like planets: their orbits would decay so that they merged with the protons in the nucleus. As this was clearly not happening in the universe we live in, it was assumed correctly that the model must be wrong.

Through the pioneering work of physicists such as Planck, Bohr and Schrödinger, a far more sophisticated model of the nature of the sub-atomic realm emerged and with it a number of counter-intuitive consequences that have caused confusion for the non-physicist ever since. One of the pioneers of quantum mechanics, Niels Bohr, even went so far as to say that, 'Anyone who is not shocked by quantum theory has not understood it.'

The problems really began in 1927 when Werner Heisenberg showed that there are limits placed upon the accuracy to which pairs of physical quantities can be measured. For example, if we try to measure the position *and* the momentum of a sub-atomic particle, the very act of measuring these quantities disturbs the particle so much that they cannot be said to have a known position and a known momentum at the same time. We can only assign *fuzzy* values for the two factors. This fuzziness is described by the *wave function* – meaning a description based entirely upon probabilities.

Now, at first glance this might seem like a trivial matter – so what if sub-atomic particles cannot be pinpointed to an exact location? But actually this is the most important concept in quantum mechanics and lies at the root of all the problems it presents for the layperson. It is also the very reason why quantum mechanics could offer the ability to build a time machine.

If we cannot define the universe at the most fundamental level, then it must mean the universe is constructed upon probabilities. There can be no certainty, no clear-cut definitions, no pure 'yeses' or 'nos'. From this spring some very strange quantum-mechanical ideas. We will consider many of them later, but for the time-travel enthusiast the most important is the fact that the universe can only be studied on a statistical level; if we probe too deeply, or try to single out individual particle transactions, we end up with nonsensical results. An alternative way to consider this is to say that the universe really only demonstrates an apparently logical framework if viewed holistically. Taking this a step further, Heisenberg himself suggested that the fuzziness of the quantum world could mean the traditional notion that *effect* must always follow *cause* might break down.

One of the consequences of this is that, given the right conditions, matter may literally 'pop out of thin air'. More specifically, according to quantum theory, it is possible to create what are called 'virtual particles' which exist only fleetingly. These are pairs of particles formed for extremely short periods of time (around 10^{-44} seconds, that is, a 1 with 44 zeros behind the decimal point). They do this by 'borrowing' energy from space, then annihilating each other and giving back the energy very, very quickly, so they never break the conservation laws of the universe.

This concept was exploited in a theory created by the Dutch physicist Hendrik Casimir in 1948. The 'Casimir effect' he described as a consequence of this theory is based upon what at first glance appears to be a simple scenario. Casimir suggested that if two metal plates are placed very close together and all the particles removed from the space they contain, what is known as a 'quantum vacuum' can be created. Because of Heisenberg's uncertainty principle, over a period of time virtual particles should appear in the 'vacuum'. These, the mathematics of quantum mechanics show, possess negative energy. Which, as we saw earlier, are precisely the sort of particles needed to open up a wormhole.

This sounds promising, but as with the other theories discussed in this chapter, the practical application of Casimir's ideas is riven with problems. The Casimir effect has been demonstrated on a miniature scale in the laboratory, but to produce enough matter with negative energy to create a workable time machine would require vast plates, some hundreds of square miles in size, which must be kept no further apart than the width of a few atoms. But in reality, it is almost impossible to produce plates of such a size, and even if this was practical it would require extraordinary technologies to make them without faults and to bring them so uniformly close together.

The use of wormholes and exotic matter with negative mass is one theoretical route to the creation of a time machine, but scientists have speculated upon other methods that in certain respects might offer at least as much hope.

The best alternative method still involves black holes and relativity, but it doesn't rely upon the possibility of wormholes.

The first mathematician to seriously speculate upon the possibility of time travel using the equations of relativity theory was Kurt Godel, a friend and colleague of Einstein's at the Institute for Advanced Study, Princeton, in 1949. Fourteen years later, a New Zealander, Doctor Roy Kerr, published a paper speculating upon the idea of a time machine using the theory of relativity as applied to black holes.

At the time, black holes were still unchristened (that honour fell to John Wheeler in 1967) but Kerr knew such objects were feasible in principle and employed the fact that time is affected by velocity and gravitational fields to demonstrate a theory of time travel. By an amusing coincidence, his paper was published on the eve of the first episode of *Doctor Who* in November 1963.*

Kerr's theory suggested that if a time machine was fired at a black hole and made to skim the edge of the gravitational well without being sucked in, time would travel far slower for the occupants of the machine. Meanwhile, the events in the world outside would be whizzing by. If the machine then travelled back to a point beyond the black hole they would find themselves in the future.

A decade later, the concept was extended by Frank Tipler from the University of Maryland, and in 1974 he detailed his ideas in a paper entitled 'Rotating Cylinders and the Possibility of Global Causality Violation', published in the highly respected journal *Physical Review*.

Tipler took things much further than Roy Kerr. In his scheme, a very advanced civilization could produce a special type of black hole called a *naked singularity*. To make this, the singularity (found at the heart of a black hole) would have to be rotating. The effect of the rotation is to twist space-time in the region near the singularity so much that time itself becomes another dimension of space through which a carefully piloted craft could be manoeuvred.

Tipler then went on to detail the design spec for the artificial naked singularity. According to his calculations you would need a cylinder 100 km long and about 10 across made of super-dense material, similar to that found in a neutron star, where all the electrons of the atoms of the substance had been fused with the protons in the nucleus. Finally the object would have to spin precisely twice every millisecond.

All of this sounds quite absurd to us today, but, say time-travel enthusiasts, we should look at the bigger picture. Such technological miracles may seem practically impossible to us, but who's to say there are not other far more advanced civilizations somewhere out there who have mastered the technology to create artificial wormholes or

---

* This was called simply *Doctor Who – Episode One* and was transmitted on 23 November 1963.

machines like the ones Tipler describes? Machines that enable them to build something like a Tardis and to travel temporally at will. It would be wise to recall that a little more than a century ago when the Wright brothers were struggling aloft in their first experimental aircraft, most people would have considered the idea of travelling to the moon an absurdity, yet only sixty-six years lay between the two events.

And yet, amazingly, Nature itself might provide the essential ingredients to build a time machine. Astronomers have found naturally occurring objects in the universe that almost fit the bill for Tipler's machines.

In 1967, the British astronomer Jocelyn Bell discovered the first pulsar. This is the remains of a dead star that has collapsed under its own gravitational field so much that the electrons orbiting the nucleus of the atoms making up the star have been jammed into the nuclei and fused with protons to form neutrons. This super-dense matter emits pulses with such regularity that pulsars are thought to be the most accurate clocks in the universe.

Since this discovery was made, special objects called *millisecond pulsars* have been observed which are so close to being Nature's time machines they may only need slight tweaking by an advanced civilization to be serviceable. Millisecond pulsars are made of material with almost the right density and they spin once every 1.5 milliseconds (one third the speed needed for Tipler's design).

So, even though we may be thousands of years away from possessing the technology to utilize such objects, the discovery of millisecond pulsars combined with the innovative ideas of Frank Tipler and others is now generating great excitement within the physics community.

These then are the possible methods of creating a device to travel through time, each based on plausible theory. The major drawback with them (aside from the technological limitations we have) is that using any of the methods we would only be able to travel back in time to the point when the device was first created. However, even if this restriction with our time machine could be overcome there remains one further serious hurdle to surmount if we are to successfully travel at ease through time. This is the problem of temporal paradoxes.

The fact that time travel can create agonizingly complex paradoxes has been known for centuries; indeed, people were thinking about such things long before any serious thought had gone into how a time machine could be designed. Today, some scientists believe that these paradox problems could actually be so severe they alone might prohibit practical time travel.

H. G. Wells set the tone with his classic novel *The Time Machine*, published just over 100 years ago in 1895. Unless Wells had a time machine himself, he would have known nothing of relativity, because the creator of the theory, the 16-year-old Albert Einstein, had just squeezed his way into a technical college in Zurich at the time and was struggling with elementary maths. Not surprisingly, the author offered little by way of explanation for his time-travel system, but he was careful not to send his hero into the past almost certainly because of the problematic paradoxes such a journey might entail.

In his short story 'All You Zombies', written in 1959, Robert Heinlein did the very opposite. Taking the bull by the horns he produced a tale involving what must be one of the most brilliant and confusing examples of a time travel paradox ever conceived.

The story centres around a character called Jane who is mysteriously abandoned at an orphanage in 1945. The child grows up with no idea who her parents are, but in 1963, at the age of 18, she falls in love with a drifter who visits the orphanage. For a while things go well, but then the drifter leaves her and Jane finds she is pregnant. The delivery of the child is difficult and she has to undergo a caesarean. Then, during the operation surgeons discover Jane has both sets of sex organs and in order to save her life they have to convert 'her' to a 'him'.

Subsequently the baby is mysteriously snatched from the hospital, Jane drops out of society and finally ends up a vagrant. Seven years later, in 1970, he stumbles into a bar and becomes friendly with the bar-tender who offers Jane a chance to avenge the drifter who had ruined her life on the condition that she joins the 'time travellers corps'. The pair then go back in time to 1963, the vagrant Jane seduces the 18-year-old female Jane at the orphanage, impregnating her before disappearing. The bar-tender then travels forward in time 9 months, snatches the baby from the hospital and deposits it at the orphanage in 1945 before dropping off Jane in 1985, where he joins the time travellers corps which has been created after the recent invention of time travel.

Jane the time traveller distinguishes himself in the corps, but after a period of dedicated service he retires to open his own bar in 1970. There he persuades a young vagrant to join the time travellers corps.

So in this tale Jane is her own mother, father and daughter. She is also the drifter and the bar-tender. But who are Jane's grandparents? She seems to be a creature out of time, self-created and totally independent of physical laws; in other words, a paradox.

There are other simpler examples of this twisting of events. Imagine a time traveller journeying 100 years into their past to the studio of a struggling artist. There he tells the artist that in the future he is world famous and recognized for a distinctive style very different from the one he is currently developing; our time traveller then proceeds to show him a catalogue of his future work. Distracting the visitor, the artist photocopies the artwork and the time traveller returns to the future. The artist then starts to copy the paintings he has photocopied.

The disturbing thing about this paradox is that it seems to offer a free lunch, and taken on face value, it breaks the laws of physics. Which came first, the paintings or the artist's fame? It also seems to cancel out the principle of free will. If beings from the future are able to manipulate the past and change our lives, where is the element of self-determination? Fortunately, there is a solution to this set of possible paradoxes utilizing a concept physicists call the 'many-universes interpretation'.

The simplest interpretation of the many-universes theory is that whenever any fundamental event occurs, the future splits into two possible outcomes or separate universes. It is easy to understand this when we refer it to our own lives. Suppose we have an important date lined up and have to travel by car to reach the agreed meeting point. Along one route we make the date, get on with our potential partner, marry and have a family. On the other, we get caught in traffic, fail to make the date and never see the other person again. This was an idea recently used very successfully in the movie *Sliding Doors*.

But this example is one on a macro-cosmic scale. According to the many-universes interpretation, every time any sub-atomic change occurs anywhere in our space-time continuum, the path splits creating two different universes. These universes may be so similar that any difference may be completely imperceptible to us. Perhaps the only variation is the position of one electron situated the other side of the universe. Even so, they will be different, and the important thing to realize here is that sub-atomic changes are happening every instant all over the universe, so the number of possible universes is almost infinite. It is because of this that the troublesome time-travel paradoxes could be written out of the equation.

One of my personal favourite arguments using this theory is to make what some may declare to be the rather sad claim that during the 1998 World Cup, a famous game between England and Argentina, which was won by Argentina after a penalty shoot-out, would, when considering the 'multiverse', have been won by England.

In the game, England were by far the better team, playing with ten men after David Beckham was sent off, and finishing the match level with Argentina, two all. The teams then entered into a penalty competition to decide the game; England lost this 4–3. My argument has always been that because England were the better team on the day, in most universes in which the game was played, England would have won, so, considering the multiverse, they were the true victors.

Consider another example. If in Universe A, 'our' universe, we go back in time and persuade our grandfather not to go on a crucial date with our future grandmother, a paradox will be avoided because, at the instant we arrive in the past, two possible futures or universes are created simultaneously, Universe A and Universe B. In Universe A, our grandfather goes out to dinner and starts a lifelong relationship completely unaware of our arrival. This leads to a future in which we are born and become a time traveller who returns to the past. In Universe B, the grandfather misses the date and we are never born. But, because we have come from Universe A, we do not suddenly cease to exist and there is no paradox.

Stephen Hawking has said that if time travel was really possible we would be visited by time tourists; but as we are clearly not, it is therefore impossible. This argument is wrong for at least three reasons. Firstly, time travellers would almost certainly be sophisticated enough to cover their tracks. Secondly, our space-time constitutes only a vanishingly small part of the entire life and volume of the universe, so it is highly probable that time travellers have not yet visited our particular time or place. Thirdly, if the many-universes interpretation is correct, only versions of ourselves in certain universes would ever be aware of the visitors. Since first proposing this idea Hawking has changed his mind and has recently declared a belief that time travel is, after all, theoretically possible.

It may be possible that one day we will be able to utilize natural aspects of the universe, such as black holes or pulsars, to develop a device to travel backwards in time. If this system is found to be impossible then we might still have the chance to use artificial wormholes. Any of these methods will require dramatic developments in physics and in particular a successful combination of quantum theory and relativity, which remains the Holy Grail of modern physics. But, on the positive side, we must also consider the fact that there might well be other exotic objects similar to pulsars and neutron stars just waiting for curious scientists to discover them. Perhaps some of these will be located nearer home.

Sceptical scientists have pooh-poohed the idea of humans ever possessing the technology to construct a workable time machine. Indeed, in May 2005, two researchers, Stephen Hsu and Roman Buniy of the University of Oregon in the United States, published findings to show that a wormhole large enough to transport a practical time capsule could never be formed, implying that this particular method is a no-no.

Time may prove this work to be correct, and indeed it could be why we have not encountered any time travellers from the future, but there is also the chance these scientists are completely wrong or that in a few years' time another team will discover a loophole to circumvent the problems presented by Hsu and Buniy.

Other scientists, though, have claimed that it is all but impossible to overcome the enormous technical difficulties of trying to manufacture a time machine. It has long been accepted that travel into the future is almost certainly an impossibility, but researchers also postulate what they call the 'chronology protection conjecture', which suggests that somehow Nature will always find a way to stop intelligent beings manipulating matter and energy to contrive the construction of a time machine to travel into the past. But then, who knows? It might well be that aliens on a planet similar to Gallifrey, currently preparing the construction of their latest model temporal transporter, are sitting back laughing at the naivety of earth scientists.

2

The Star Fields of Gallifrey

Is There Life on Other Worlds?

Attempt the end, and never stand to doubt;
Nothing's so hard, but search will find out.
Seventeenth-century poet Robert Herrick

In the universe of the Time Lords there is certainly no shortage of alien life forms. Indeed, the central premise of *Doctor Who* is that the eponymous doctor, seen as a renegade by the authorities on Gallifrey, has become oddly attached to the human race and protects it (contrary to the rules of the Time Lords) from the destructive wishes of alien races. This plot construct is of course a staple of science fiction, from *The War of the Worlds* to *Star Trek*. Indeed, the majority of orthodox science fiction written during the past century is set in a universe in which there are many populated planets, and alien races travel freely between the stars.

Sadly, in the real world the very opposite holds, in that there is still no hard evidence of life existing beyond the limits of earth's atmosphere. But of course, we may wonder, we may speculate, and we may apply science, to look objectively at the subject of life on other worlds. We need to do this because the question of whether or not there is life on other planets is one of the most profound we can ask, and, for many, finding an answer is an obsession.

We know there is certainly life on one planet – the Earth. But with just this one example upon which to build hypotheses, knowing for sure whether the series of events leading to life here is unique or extremely common is impossible. We need evidence. And, because of the almost unimaginable distances between stars, it is only now we can travel outside our own planetary atmosphere and develop machines to see into the deepest recesses of space that we may hope for some answers.

It has been known for at least a century that we are unlikely to find life on any other planet in our Solar System. The reason for this is that certain basic requirements are needed to allow life to form. One of these is the need for water, because no biological processes can take place without it. This means that the range of 'allowed' temperatures and pressures for any life-supporting planet is limited. Another essential characteristic for any life-supporting world is that it must possess an atmosphere, not only to allow respiration but to ensure that water remains on (or close to) the surface rather than evaporating into space.

Some extremely hardy creatures can survive in the most inhospitable conditions – simple multicellular organisms have been found in natural hot-water vents in the depths of the ocean, and some particularly tough bacteria have been shown to survive in nuclear waste. But even these seemingly extreme conditions are not really that extreme when we consider the environments of the planets we know most about – those in our solar system.

Of the two planets nearest to the Sun, Mercury and Venus, Mercury has extremes of temperature with part of its surface an inferno and the other a frozen wasteland, while Venus has surface temperatures of some 500 °C (800 K). Beyond the third planet, Earth, lies Mars, thought by many to be the most likely contender for extraterrestrial life within our Solar System. In fact, as recently as 1877 the astron-, omer Giovanni Schiaparelli created great excitement by announcing that he had observed a network of what he called *canali* on the surface of the planet. *Canali* was wrongly translated into English as 'canal' instead of its true meaning, 'channel', and astronomers all over the world began to see increasingly complex canal systems as the rumours spread. Sadly, although the news inspired H. G. Wells to write *The War of the Worlds*, there are no canals on Mars: the effect is produced by a natural coloration of the surface.

The *Viking* probes of the 1970s found no trace of life on Mars – not a single microbe. More recently, NASA's series of Mars landers, especially *Spirit* and *Opportunity* which landed in 2004, have offered fresh hope by finding clear evidence of frozen water at the poles of the planet. This shows that even if life no longer exists on Mars, it may have once flourished there.

Beyond Mars lie the gas giants, Jupiter and Saturn, with atmospheres containing toxic gases constantly churned up within powerful vortices. The two largest satellites of the Solar System, Titan, orbiting Saturn, and Ganymede, Jupiter's largest moon, might be more promising candidates for harbouring life. The *Voyager I* probe of the 1980s passed close by Titan and found what are believed to be organic molecules on its surface. More recently, the *Huygens* probe, which actually landed on the surface of Titan in January 2005, found lakes of liquid and running rivers. These, though, were composed not of water but of the organic molecules methane and ethane. Today, Titan has a surface temperature of around –150 °C (123 K) and a toxic atmosphere with little trace of oxygen, so the chances of these molecules developing into living matter is very small. However, in the distant past conditions may have been more suitable for the development of simple life forms.

Ganymede, the largest moon in the Solar System (bigger than both of the planets Mercury and Pluto), possesses a strong magnetic field, a rare property shared only with the Earth and Mercury.* Probes have recently found traces of ozone in the atmosphere of Ganymede, which for the exobiologists (as those who study extraterrestrial biology are known) is another indicator that some form of life might once have thrived on this moon.

At the outer edge of the Solar System, Uranus, Neptune and Pluto offer little comfort for those searching for what scientists refer to as carbon-based life forms (we'll return to the significance of carbon later in the chapter), because again either the temperatures are too low, their atmospheres too noxious or, in the case of Uranus, the entire planet is a single ocean of superheated water, warmed by volcanic action and covered in poisonous gases.

Because there appears to be a very narrow spectrum of conditions

---

* Although Mars and the Jovian moon Io may soon prove also to be members of this club.

which could support life, to find it, particularly life we can readily recognize, we must turn our thoughts and our telescopes and probes to regions beyond the tiny confines of our Solar System, to the distant stars. But, the problem we face then is precisely that distance. Our Solar System is vast by everyday measures, some twelve billion kilometres across, but this becomes a meaningless speck and such numbers trivial when we begin to imagine contacting beings living on planets orbiting other stars.

The nearest star to Earth other than our own Sun is Proxima Centauri, which lies 4.3 light years from Earth. What this means is that light from this star, which as we know from Chapter 1 travels at just over 300,000 kilometres per second, takes 4.3 years to get here. This equates to a distance of 300,000 × the number of seconds in 1 hour (3600) × the number of hours in 1 day (24) × the number of days in 1 year (365) × 4.3, which comes to a little under 4×10^{13} kilometres (4 with 13 noughts after it, or 40 million million kilometres). This is roughly equal to 160 million trips to the Moon. At the speed *Apollo* capsules travelled (about 40,000 kilometres per hour) it would take 100,000 years to cover the distance to even this, merely our nearest neighbour.

So, until we develop the technology to improve our speed, we can only hope to a) contact aliens using light-speed signals such as radio waves, b) wait for them to contact us, or c) use telescopes and other technologies to discover as much as possible about other planets we may find orbiting nearby stars.

But just what are the chances of there being intelligent life beyond our own world?

On this matter scientific opinion is split. There are those, like the astronomer Frank Drake, creator of the first SETI (Search for Extraterrestrial Intelligence) project, who believes the universe is teeming with life. At the other end of the spectrum, writers and pundits such as Marshall Savage, author of the *Millennial Project*, and the physicist Frank Tipler think we are totally alone.

The problem with trying to come up with any form of definitive answer or even an approximation is that we have no clear idea of all the variables to be considered or how these inter-relate. For example: how likely is it that molecules of DNA can form given a long enough time period? How frequently do planets form around stars? How likely is it that even complex molecules can evolve into living material? We know all these things have happened at least once, here, but has it been *only* once, or billions of times?

To try to quantify the argument, in 1961 the pioneer in the search for extraterrestrial intelligence, Frank Drake, produced a now-famous formula which has since become known as the 'Drake Equation'. It is very straightforward and a potentially powerful tool for the astronomer, except that almost all the variables can show a range of values and no one is yet sure what numbers to put in. It is the work of astronomers, biologists and geologists to gradually narrow down each of these variables to something more workable, and to then come up with some form of solution to the Drake Equation.

The equation is:

$$N = R \times f_p \times n_e \times f_l \times f_i \times f_c \times L$$

Although this might look daunting, to use it is actually as easy as working out your travel expenses. The letter N signifies the number of civilizations in our galaxy trying to make contact. The symbols on the right hand side of the equation represent separate factors which have to be considered in addressing the question: is there life beyond Earth? (Each term is considered in isolation; in other words the number assigned to say f_p is independent of that given to L, f_i or any of the others.) When numbers for all of these factors are plugged in, we end up with a figure for N.

So what are these factors?

Let's first consider R. This stands for the *average rate of star formation*. A common misconception is that the universe was made at the time of the Big Bang and that was that – no change ever since. Of course, this is not the case. The prevailing theory is that the universe is expanding and stars and planets are being created and dying constantly. Scientists are beginning to actually see this birth process using instruments such as the Hubble Space Telescope. It seems that some parts of the galaxy are more fertile than others and the process of star birth is far slower than it was at distant points in our galaxy's past, but by a conservative estimate, astronomers think that about ten new stars are formed in our galaxy every year. So **R** is one of the variables which is pretty much agreed upon: 10.

f_p is *the fraction of stars that are good*. By 'good', astronomers mean suitable for forming and keeping stable, Earth-like planets in orbit around them as part of a solar system. This constitutes quite a refinement. The age of the star must fall into a certain range. If it is too old, its fuel will be running down and it will emit radiation that hinders the formation and survival of carbon-based life. Also, as a star ages, the rotation of planets in orbit around it begins to slow. If the star is more than around six billion years old (our Sun is about five billion years old), this will have a dramatic effect: planets orbiting very old stars will have stopped rotating altogether and will have one face permanently turned towards their sun while the other hemisphere remains in permanent night. If the star is too young, it may not have had time to allow planet formation and the mechanism that creates and evolves life forms to run its course.

More importantly, planets that can sustain life forms capable of developing civilizations could not be found orbiting pulsars or quasars – exotic stellar objects that emit forms of damaging radiation and make the home star unstable over long time periods.

Finally, many stars are binary – that is, they are made up of two stars orbiting each other. Such systems can possess planets, but binary stars are generally considered less likely to offer stable conditions such as those provided by our own Sun.

When Drake first proposed his equation, the value for f_p could only be guessed at, but recently astronomical findings have begun to narrow down the possible range of numbers. Back in the early 1960s, Drake placed f_p at about 0.5, in other words, half the number of stars in the galaxy were potentially able to form planets. But when observational techniques improved and new data was gathered, this figure was shown to be rather optimistic.

Using present-day technology, far off planets cannot be seen in the way astronomers observe the planets in our own Solar System; the distances, as I have mentioned, are simply too great. It has been estimated that a telescope the size of the moon would be needed to observe clouds and coastlines on a planet fifty light years from Earth. And distance is not the only problem. Imagine trying to detect the presence of a firefly perched on the edge of a spotlight from a few hundred miles away. The light from the spotlight would completely swamp the image of the firefly. In the same way, the light from an orbiting planet (which is merely reflecting light from its sun) would be totally overwhelmed by the far greater brilliance of the star.

These two problems would, you might think, stop us ever knowing if any star other than our own possesses planets, but there are other ways of knowing if a planet orbits a distant star.

The best technique we have today is observing 'wobble'. Picture a hammer thrower at the Olympic Games, spinning on the throwing circle and just about to let go of the hammer. The athlete has a great pull on the chain and the hammer, but the hammer (which weighs about 7 kg) also has a pull on the thrower, who might weigh around twenty times as much. With suitable instruments this pull, or 'wobble', could be measured. In the same way, a planet in orbit around a star will exert a pull on the star – a much weaker pull than the star's on the planet, of course. Obviously, the bigger the planet, the greater the effect. The technique for analysing this attraction involves the measurement of 'radial velocity' and the taking of 'Doppler measurements'.

Although this is a method requiring very sensitive instruments – the difficulty in measuring the wobble of a star has been compared to using a telescope on Earth to see a man waving on the moon – it has produced some hugely important results. And although Doppler measurements remain the most widely used system and account for some 90 per cent of the finds, during the past decade a huge range of other techniques (including photometry and microlensing) have been developed which the modern astronomer uses to detect planets of different types.

The discovery of a planet orbiting the star known as 51 Peg in the constellation of Pegasus, announced in October 1995 at a conference in Florence, startled the astronomy community and made headlines around the world. The planet (known simply as 51 Peg b) is about half the mass of Jupiter (the largest planet in our Solar System and about 300 times larger than the Earth), and it orbits 51 Peg at a distance of about eight million kilometres.

Although most astronomers believe it unlikely that gas giants like

Jupiter could sustain carbon-based life, it is suspected that for a solar system to have any chance of containing an Earth-like planet it needs at least one Jupiter-type planet, which would probably be found further away from the star than the region in which solid, cooler planets would be located. The reason for this is that gas giants act like vacuum cleaners, soaking up stray asteroids, comets and meteors that enter the system. In this way they would protect the inner, Earth-like planets, increasing the chance of a stable environment within which life could form and a civilization develop.

However, the planet found around 51 Peg is in the wrong place. It is far too close to the star (Mercury orbits the Sun at a distance of 46–70 million kilometres), and it circles the star too rapidly, taking only four days to complete an orbit (Mercury takes eighty-eight days). But, it *is* a planet, and the discovery created a revolution in our way of thinking about the cosmos. From October 1995 we have known that our Solar System is definitely not the only one.

Within months of the discovery of 51 Peg b, more solar systems were discovered. In January 1996, two new planets were found around different stars, one orbiting 70 Virginis in the constellation of Virgo and the other in the constellation of Ursa Major, orbiting a star named 47 UMa. Both of these are around thirty-five light years from Earth; they are Jupiter-types and, in a similar way to the gas giants of our own solar system, they may be found in orbits far from their suns.

Since the first breakthrough in this field over a decade ago, this area of research has mushroomed incredibly. At the latest count, 155 planets and 136 planetary systems have been found, fourteen of these are solar systems in which two or more planets have been detected, and many of these systems contain planets much smaller than Jupiter. One of the most interesting is a planet recently found orbiting the star Gliese 876, just fifteen light years from Earth. It is less than six times the size of our planet and early indications show that it might be rocky rather than gaseous. Unfortunately, it is not really a contender as a life-bearing world because its orbit lies only two million miles from its sun, making the surface temperatures almost certainly too hot to sustain life.

One of the newly discovered solar systems most favoured as a place that might support life is the solar system of 55 Cancri, a star that is almost identical in size to our Sun and lies some forty light years away from Earth. Scientists have so far found four planets orbiting this star, including two gas giants and a planet one-twentieth the size of Jupiter. This system is now viewed as one of the best candidates for extraterrestrial life and has prompted considerable interest within the astronomy community.

So, what do these findings mean for the number we assign to f_p? Well, although these recent finds have greatly encouraged astronomers, f_p remains a very difficult value to quantify. Only a very few of the new planets are significantly smaller than a gas giant, since it is very difficult to detect small planets using the 'wobble' method. Viewing the matter conservatively, most scientists believe the initial guess for the value of f_p was too high. Instead of 0.5, a more realistic estimate might now be 0.1, indicating that one in ten stars are capable of developing and sustaining a planetary system.

Next we come to the term n_e, which is *the number of planets per star that are Earth-like*. Once again, we are bound by very limited experience. In our Solar System, there is really only one Earth-like planet. Mars may once have possessed a more hospitable atmosphere than it does today, but now the surface temperature ranges from –50 °C (223 K) to 0 °C (273 K) and the atmosphere is so thin (with an atmospheric pressure about one hundredth of Earth's) that humans would need to carry their own oxygen supply to work or to move around on the surface. At the other extreme, Venus has an atmosphere made up almost entirely of carbon dioxide (CO_2), which produces such a severe greenhouse effect that the Venusian surface is hotter than that of Mercury.

So, again being conservative, let us assume our solar system is typical of one in which life might be found and assign n_e the value 1.

f_l in the Drake Equation stands for *the fraction of Earth-like planets upon which life could develop*, and in trying to assign a number for this parameter we really do enter into almost totally uncharted territory.

Firstly, what is life? It might seem an obvious question, but biologists are quick to point out that the standard ways in which to define life are nebulous. Living beings grow and move, but so do crystals:

they produce regular patterns made of repeated simple units in much the same way cells do. Furthermore, totally inanimate water or any other liquid may flow, or move. Life forms use energy, but so do computers, trains, rockets. Perhaps a better definition would be to say that a living being can *control* energy.

An alternative could be to argue that only living things process and store information, but this after all is the sole purpose of a computer. The debate about the possible future development of intelligent computers still rages, yet the desktop computer I'm using to write this could certainly not be described as living.

So, what other criteria could we use? Could the ability to reproduce constitute life? The problem with this definition is that a flame reproduces. Refining this description, though, gives us what is probably the best definition of life, which is that all life forms reproduce and pass on genetic material (inherited characteristics) to their offspring.

So much for definitions, but in this investigation I'm really interested in reaching a conclusion about *intelligent* life forms with which we can communicate in some manner. So, we need to refine things a little more. It may be that any number of exotic creatures live in this almost infinite universe, but if they are very different from us then the chance of contacting them or communicating with them is reduced. There is even the possibility that we have encountered such beings and have, for one reason or another, been totally unaware of them, or they of us. So, to allow for the condition of 'life as we know it' we need to think in terms of carbon-based life able to communicate with us.

But why is the caveat 'carbon-based' important?

According to the laws of physics (which in turn give us the rules of chemistry and subsequently biology), carbon is the only element which can form complex molecules, known as organic molecules. For 'life as we know it' to have developed elsewhere, such alien beings must naturally operate within the same narrow limitations of temperature, pressure and radiation as we do, not least because that would be the only way we could communicate with them. Within those parameters the only element to form the organic molecules that make cells, tissue and flesh is carbon. Carbon has the unique ability to form very strong inter-atomic bonds with other elements such as nitrogen, oxygen and hydrogen. As well as this it can form multiple bonds with other atoms of carbon. This allows it to create a vast range of molecules, some of which may contain tens of thousands of atoms. No other element can approach this level of versatility.

For a planet to be the home of carbon-based life it must have possessed a limited set of environmental conditions and materials in its primeval history and a subsequent set of finely tuned conditions and materials to allow that life to have evolved and flourish. Sceptics argue that these conditions are unlikely to be duplicated in the universe and that the probability of life evolving is therefore slight, but there is a growing body of evidence to oppose this view.

In 1953, just as the structure of DNA was being elucidated by Crick and Watson in Cambridge, two scientists at the University of Chicago, Stanley Miller and Harold Urey, were investigating the initial conditions on Earth that produced the biochemical environment in which life began. It was known that life originated on Earth a little under four billion years ago and that the predominant chemicals in the atmosphere at that time were ammonia, water and methane. Miller and Urey placed these chemicals in a jar and allowed an electrical discharge to pass through the mixture. After sustaining this process for several days they found a red-brown deposit had formed at the bottom of the jar. When this 'primeval soup' was analysed the researchers found it contained amino acids – organic molecules that provide the essential building blocks of all life on Earth.

Further experiments showed that a wide range of molecules essential for life could be formed in this way. Miller and Urey then added another simple molecule, hydrogen cyanide (HCN), found in volcanic gases, and discovered that a number of complex molecules which are key to the formation of proteins and DNA were formed. Although these molecules are a long way from the elaborate molecules that encode the production of proteins and the day-to-day workings of cells – the giant structures of DNA (deoxyribonucleic acid) and RNA (ribonucleic acid) – Miller and Urey postulated that a 'soup' of life-forming molecules could have been brewed in the Earth's atmosphere during the space of just a few years. Recently, Stanley Miller has declared that these molecules could have developed in complexity and produced living cells within perhaps as little as 10,000

31

years. Flying in the face of critics who claim life on Earth is unique, Miller is sure that, based on his own experiments, given the correct environmental conditions and the proper blend of chemicals life could form on any planet.

Finally, the fossil record shows us that life began on Earth at the earliest opportunity. In 1980, fossils of creatures called 'stromatolites' or 'living rocks', the simplest and probably the most ancient form of life on Earth, living over 3.5 billion years ago, were found in the Australian desert. We know the environmental conditions for life only became suitable less than four billion years ago, and so it would seem that a period perhaps of only a few hundred million years passed before the very simplest life forms began to appear. This does not provide evidence for the evolution of life on planets other than the Earth, but it illustrates the notion that, given the correct conditions, life will appear readily.

The great proponent of extraterrestrial life, the late Carl Sagan, once wrote: 'The available evidence strongly suggests that the origin of life should occur given the initial conditions and a billion years of evolutionary time. The origin of life on suitable planets seems built into the chemistry of the universe.'

What Sagan was describing here is the principle of 'self-organization'. In recent years it has been suggested that certain physical and chemical systems can leap spontaneously from relatively simple states to ones of greater complexity or organization. This organizing principle, some argue, is a form of 'anti-entropy' effect which may be, in some mysterious way, linked to life itself. As we saw in Chapter 1, entropy is the 'level of disorder' in a system, and in Nature entropy always increases – an apple left to stand will gradually decompose, its cells break down, and the 'neat', 'organized' form of the fresh apple will decay into a disorganized pool of liquid. The self-organization principle could, it is believed, help to reverse the natural tendency for entropy to increase in the universe at least on a local level. As a consequence, the chance of life deriving from a collection of complex organic molecules is greatly increased.

According to Frank Drake, 'Where life could appear, it would appear,' and he boldly assigned a value of 1 to the parameter f_l. In other words, there is a 100 per cent chance that a suitable planet will form life if it has the correct conditions. Others, such as the Nobel Prize-winning chemist Melvin Calvin and Carl Sagan, have concurred, believing life is more likely than not to form on a suitable planet. For others, those who do not believe in the existence of extraterrestrial life, the value for f_l is the most crucial of all the terms in the Drake Equation. They place it at 0, which would consequently make N equal to 0 – no life anywhere else in the universe. The value for f_l probably cannot be anything other than 1 or 0, so, for the purposes of this discourse, I will give it a value of 1.

Next, let's consider the value for f_i. This is the term that refers to *the fraction of Earth-like planets where life has developed intelligence*. Again, when we first contemplate this expression, we are struck by the need to define. This time the question is: what constitutes intelligent life?

Many would argue that dolphins and whales are highly intelligent animals that could have formed a civilization if they had evolved on land. They can communicate with members of their own species and have been known to interact in a highly intelligent fashion with humans. Attempts have even been made to decipher the complex sequence of clicks and squeaks these creatures use to communicate with one another. In a different sense, ants and bees act in an intelligent way when they are viewed as a collective of individuals each acting as a unit in a larger society, a *gestalt*. So, if we were to apply Drake's Equation to our planet, we could arrive at a value for f_i between 1 and at least 4, but again, being conservative, let us take the value of 1, representing the human race.

The penultimate term, f_c, represents *the number of intelligent species who would want to communicate with us*, and once again we are faced with a dilemma in fixing a value for this parameter. In order to use this term we have to place some limitations upon how we arrive at a value. We must first assume that an intelligent race uses some form of electromagnetic radiation with which to communicate and to interact with the universe. Most scientists would agree that it would be unlikely for an intelligent species to have developed without using any form of electromagnetic radiation. An alien intelligence may utilize extreme regions of the spectrum, they may see in the infrared or the ultraviolet because of the nature of the light emitted by their sun; alternatively, they may live in extreme conditions such that vision is as unimportant to them as it is to some deep-sea creatures. But whatever unusual situation we may consider, all alien races must use some form of electromagnetic radiation. If they do not, then such beings would fall out of the category of 'life as we know it'.

As a civilization, we employ a wide range of radiation, from radio and television signals to X-rays, from ultrasound to microwaves, so it is likely that any civilization at least as advanced as us would also employ similar electromagnetic waves; they may even have developed something similar to television or radio. Of course, they may not have created any form of entertainment system which leaked signals into space as our televisions have done in recent decades, but if they were actively interested in communicating, they should be able to build equipment that would receive and decipher signals from space.

This then leads to a question concerning the sociological and psychological make-up of an alien intelligence: would they necessarily want to communicate with us? It is a serious point that the signals we have been sending inadvertently into space may have presented our race in a very poor light. For some seventy years our calling card has been television signals conveying images of everything from the most violent Hollywood films to news coverage of war, famine and torture. These images have leaked into space in far greater quantities than any form of contrived, politically correct message we may wish to send to our celestial neighbours. Many of these signals would be too weak to reach distant stars, but this is perhaps underestimating the sensitivity of alien detection systems. Television signals are no different to any other forms of electromagnetic radiation in that they travel at the speed of light. It is therefore conceivable that alien civilizations living on planets up to seventy light years away could be chuckling at our antics, or perhaps battening down the hatches for fear we'll ruin the neighbourhood.

So, what value do we give f_c? On the one hand, it would seem likely that any civilization would eventually develop a form of long-distance receiving and transmitting system using electromagnetic radiation to enable communication; but how many races would want to make contact? There could be an abundance of races busily communicating with one another but excluding us. Equally, alien civilizations could prefer to keep themselves to themselves irrespective of whether or not they have been warned off. Weighing up these factors, a conservative estimate would be that 10–20 per cent of intelligent aliens would want to communicate, so f_c would be, say, 0.1.

Finally, we come to L, which represents *the lifetime of a civilization* (in years). And again, in attempting to assign a value, we face another complex series of permutations. The value of L puts us out on a limb, for we have to consider hypothetical sociological factors for a hypothetical race, but again we do have one example to draw upon – our own experience.

There is perhaps some synchronicity in the fact that our race developed weapons of mass destruction at almost the exact point we revealed ourselves to the universe with our electromagnetic signals, and it could be that many races are destroyed at the very point they could make contact with their neighbours.

Since Frank Drake first devised his formula in 1961, there has been much debate amongst scientists about the value of L, and during the past forty-five years the political and social Zeitgeist has altered radically. The Cold War has ended, but the threat of nuclear destruction is still very much with us, and the killer instinct of humankind has not changed in the slightest. Perhaps the ease with which human beings make war is irreducibly linked with our drive to progress and advance. It is possible that the instinct that drives us to communicate derives from the same source as our aggression. If this is the case, other species may be no different; such correspondences may even be universal. If so, it would be likely that a large proportion of civilizations do destroy themselves at around the time they develop the technology to communicate beyond their own world.

There are also a number of other ways in which a species based on a single planet may become extinct. Scientists are only now beginning to realize the very real danger of comets or asteroids colliding with our planet. It is believed that a devastating asteroid collision caused a sudden traumatic alteration in the ecosystem of the Earth some sixty-five million years ago, resulting in the extinction of the dinosaurs, and there have been a number of documented near-Earth collisions during the past century. The massive explosion reported in

Siberia in 1908 which devastated hundreds of square kilometres of forest in the region of Tunguska is thought to have been caused by a meteorite exploding several kilometres above the ground. If this had occurred over London, millions would have died. An object only a few times larger than the Tunguska fireball colliding with the Earth would not only devastate a wider area, but the dust thrown up by the collision could produce a blanket around the entire planet capable of destroying all life on the surface. If it landed in the sea, the tidal effects would be almost as damaging.

There is also the question of planetary resources. We as a race are perilously close to over-exploiting the resources of our world, and we are already capable of severely damaging planetary mechanisms that are there to maintain an ecological balance. It is conceivable that other civilizations have followed the same path and gone further, completely destroying their own environments. Such threats as reduced fertility, AIDS, superbugs and nuclear terrorism are all further potential civilization-killers.

One possible conclusion to be drawn from these considerations is that civilizations either survive little more than 1,000–2,000 years or else they continue for perhaps hundreds of millennia. It is possible that many races pass through a 'danger zone' during which they have a high chance of destroying themselves, but if they come through it, they develop into highly advanced cultures capable of interstellar travel and galactic colonization.

The value assigned to L may also depend upon astronomical factors. If we assume that life may form on a large number of planets and that those life forms could evolve into intelligent, civilized beings, the time at which life began on their world would be a crucial consideration.

The universe is believed to be approximately fourteen billion years old, and our sun is a very typical star located some two thirds of the way along one of the spiral arms of the Milky Way galaxy – itself an 'ordinary' galaxy among an estimated 100 billion others. In astronomical and geological terms, the Earth is quite average, and life began to appear here four billion years ago, or around ten billion years after the Big Bang. But it is quite conceivable that a great many planets around other older stars cooled long before our own planet.

Astronomers have observed the death of stars far more ancient than our own. If any planet around these stars had brought forth life, a civilization that formed there would either be ancient interstellar voyagers or long dead.

To find a sensible value for L, we must assume a normal distribution of ages for successful civilizations. If L for a particular planet is 2,000, the race may possibly have destroyed itself long ago and is of no further interest. But L may be much larger. It is possible there have been and still are civilizations hundreds of thousands of years old. Equally there may be a large number of very young civilizations, perhaps no more than 2,000–3,000 years old. Most civilizations that have survived and are able to communicate would be somewhere between the two extremes.

Drake and his colleagues have placed a value of 100,000 on L. This seems rather arbitrary but nevertheless, if we put any figure over the 2,000-year watershed the equation gives us a correspondingly large number for N, the number of advanced civilizations wanting to make contact.

To recap, we have assumed $R = 10$, $f_p = 0.1$, $n_e = 1$, $f_l = 1$, $f_i = 1$, $f_c = 0.1$, L = a large number. Putting these figures into the Drake Equation, we reach a very simple conclusion.

$$N = 10 \times 0.1 \times 1 \times 1 \times 1 \times 0.1 \times \text{a large number}$$

The 10 and the first 0.1 cancel out, giving us:

$$N = 1 \times 1 \times 1 \times 1 \times 0.1 \times \text{a large number}$$

If we call L (the average age of a civilization) 100,000, it means there are 10,000 civilizations sharing just this single galaxy (one of 100 billion galaxies, remember). Frank Drake believes the value of L to be much greater, which would mean that N would be correspondingly larger. In fact, some enthusiasts suggest that all the numbers postulated here are far too conservative and that N is more likely to be a number in the region of tens of millions, or even hundreds of millions. This may seem overly optimistic, but when we consider that our galaxy contains upwards of 100 billion stars, an estimate of 100 million civilizations means that there is still only one such race for every 1,000 stars.

So, how likely is it that we will communicate with one of these civilizations? So far every attempt to make contact has failed; but that has not been for want of trying. The first serious scientific suggestion for how best to detect signals from alien civilizations came in a paper published in the science journal *Nature* in 1959. It was written by the Italian astronomer Giuseppe Cocconi and the American physicist Philip Morrison, who reasoned that if aliens wanted other species to contact them, they would make it as easy as possible for them to do so. What they meant by this was that an alien intelligence would broadcast a signal that would have universal meaning and be within a range detectable via radio telescopes.

Cocconi and Morrison chose as their standard the frequency 1.42 GHz (gigahertz). The reason for this choice is that 1.42 GHz is the frequency at which the element hydrogen is known to resonate. Because hydrogen is by far the most common element in the universe, it was assumed an alien intelligence which had developed radio technology would know the resonance frequency of the element and expect a recipient to be equally knowledgeable.

Enthusiasts immediately began scanning the sky, calibrating their radio telescopes to this frequency. The first SETI project was led in the early days by Frank Drake, who in 1960 established a team at Green Bank, West Virginia, in the United States. This was followed soon after by another SETI project created at a giant radio telescope centre called Big Ear, in Ohio, directed by astronomer Bob Dixon, who, like Drake, has been searching the skies his entire career.

Today there are a number of SETI projects running concurrently around the world. In 1992, NASA created a programme with an annual budget of $10 million. Their approach was different from the early experimenters. Whereas Drake and others had focused on the frequency of 1.42 GHz and a select group of prime candidate stars within fifty light years of Earth, NASA decided to blitz the heavens with a broad sweep, taking in as many likely candidate stars as they could and selecting a wide range of frequencies. Sadly, we will never know if the project would have garnered conclusive results because, little more than a year after it was established, funding was terminated by the interference of congressmen who thought they were paying for the indulgences of UFO cranks.

Such ignorance does nothing to help the serious search for life beyond our planet, but luckily there are others who have a more open-minded approach. Rather aptly, Steven Spielberg, director of *ET* and *Close Encounters of the Third Kind*, is funding a project on the east coast of the United States, and other wealthy enthusiasts are putting money into the search at a variety of sites around the world. While the original NASA SETI project was scrapped, three years later it was picked up by others using private funding. One of the most determined groups was that behind Project Phoenix, a pan-global operation which used the world's largest radio telescopes to sweep the universe between 1 and 10 GHz. NASA have also developed what they call spectrum analysers to search across wide ranges of frequencies, and they are currently devising software to filter out noise and other interference in the signal.

The recent flurry of interest prompted by the discovery of hundreds of extrasolar planets has inspired a more direct approach to contacting alien life if it is out there. In 2003, a new project, Cosmic Call, created by a group called Team Encounter, began transmitting a narrow beam signal directly towards a shortlist of pre-selected stars. These are stars which possess more than one planet and especially those which have a high chance of having Earth-like planets in orbit around them. The most favoured candidate is 55 Cancri, but the list drawn up by Team Encounter includes four others: 7 Ursa Majoris, HIP 26335 (located in the constellation of Orion), HIP 7918 (in Andromeda), and HIP 4872 (found in the constellation of Cassiopea).

This group of enthusiasts, which includes exobiologists, the first female astronaut, Sally Ride, and the rock musician Greg Lake, use a 230-foot diameter radio astronomy dish (one of the world's largest steerable radio astronomy dishes) at the Evpatriya Radio Astronomy Facility in the Ukraine, and the signal is designed to be easily decoded. The transmitted message includes information about what scientists on Earth know of the physical universe – including many physical constants, such as c, the speed of light – along with mathematical universals such as the largest known prime number and the current number of known decimals of Pi. According to Yvan Dutil, a Canadian astrophysicist involved in the message conception, 'The message is constructed to induce a reply. We show we are curious and ask some questions.'

Yet, despite the many attempts to use advanced technology to detect life on other planets, after almost fifty years of searching no conclusive evidence for what have been dubbed LGMs (Little Green Men) has been found. Nevertheless, the search goes on, and with the recent confirmation that many stars possess planets, the thunder has been taken out of cynical doubters such as the congressmen who saw no value in the funding of SETI. If there is life on other planets, which would seem more likely than not, we do have a chance of contacting them one day. If only one in every thousand stars has planets where an advanced civilization has developed and has survived, there will be at least one such star within fifty light years of Earth. Perhaps we are sending signals at the wrong frequency and tuning our instruments to the wrong region for that particular civilization. Maybe we are unlucky enough to be neighbours to a race who either do not want to be contacted or have developed a technology that is so different from ours that we cannot yet reach one another. It could also be that signals sent from more distant planets have not yet reached here.

Sceptics quote what has become known as Fermi's Paradox after the famous Italian physicist Enrico Fermi, who declared in 1943 that if aliens existed they would be here. This, though, is not only a tautological argument, it demonstrates staggering arrogance, for it assumes that it is a simple matter to detect alien life, and that we are so important that aliens would want nothing more than to contact us.

Whatever the future efforts of scientists to contact alien beings brings, the effect of such a discovery upon our society and the mindset of each of us would be enormous. As the physicist Paul Davies has said: 'There is little doubt that even the discovery of a single extraterrestrial microbe, if it should be shown to have evolved independently of life on Earth, would drastically alter our world view and change our society as profoundly as the Copernican and Darwinian revolution.'

Perhaps in the not-too-distant future this happy event will be realized, constituting a discovery that would truly change our view of ourselves and the universe in which we live. Until then, the intriguing question of whether or not there is life out there will at least remain exciting enough for people to continue looking for answers. And meanwhile, science-fiction writers will continue to let their imaginations wander.

3

Bring on the Daleks

What Would Aliens Really Look Like?

There is no trusting to appearances.

Richard Brinsley Sheridan

Okay, so you either believe there is life on other planets or you don't. Unfortunately, if you do believe the universe is teeming with life, there is no firm evidence to support your argument. However, as the Astronomer Royal, Martin Rees, once declared: 'Absence of evidence is not evidence of absence.'

I will stand up and be counted proudly as a believer in extraterrestrial life. To me it seems ludicrous that in this vast universe we would be the only intelligent beings, and it strikes me as supreme egotism to believe this. As a consequence, I've always been interested in speculation about what alien beings might look like, and in recent decades others with this interest have made inroads into our understanding of the possibilities of alien biology, a study now known as exobiology.

What we know of alien life is of course based upon pure conjecture, but we may be guided by the laws of science that we apply regularly to more prosaic matters. The only model we have is life on Earth, for this is the only world we are sure harbours life. This is of course a limitation, and unfortunately we will not know whether the theories of the exobiologists are even close to reality until we encounter other life in the universe. To cover this subject we need to ask the questions exobiologists ask and try to formulate sensible answers based upon the very latest information available. To do this we will need to bring together a collection of disparate disciplines, just as exobiologists do, and try to evaluate ideas to see whether they are 'right' or 'wrong'.

Sadly, we will not be able to reach any definite conclusions, just probabilities based upon what is presently known.

The first question we need to ask is the most fundamental of all: what is life?

I mentioned this in the last chapter and suggested that a good definition might be that *all living things reproduce and pass on genetically inherited characteristics*. Fleshing this out a little, it would be better to say that all living entities pass on inherited characteristics to their offspring and that this material undergoes some degree of mutation. In other words, the living thing in question has taken part in the evolutionary process via natural selection, they haven't simply produced exact copies of themselves. But perhaps the final word on the matter should go to Carl Sagan who, shortly before his death, defined life as: '. . . any system capable of reproduction, mutation and the reproduction of its mutations'. What he meant by this was that life was represented by any entity that allows for variation from generation to generation using the mechanism of evolution via natural selection, an entity that can pass on its characteristics, reshuffled by the processes of reproduction so that those characteristics will not appear to be exactly the same in the next generation.

But even this is actually not a totally satisfactory definition, because there is the problem for the argument created by clones, which are produced without sexual reproduction and which are perfect copies of their parents. Dolly the sheep looked very much alive, but she would not have fitted the above definition of life.

So, if we want to move on in this exploration we have to accept a definition which links life with the ability to reproduce via a mechanism that allows for mutation. This is because the only way we know that life can evolve is via this route. Dolly the sheep may have been a fully functioning sheep that could do anything any other sheep can do (apart from reproduce), but she and any descendants of hers produced via cloning play no role in the evolutionary development of the species of which she was a member.

So, it is clear that evolution and life are linked. Indeed, any biologist would support the idea that without evolution there can be no life. So, whether that life is on Earth or a planet orbiting 51 Peg, evolution will be a fundamental process steering life there.

So, how would evolution work on an alien planet?

Well, we cannot be sure, but it would seem likely it would operate in the way it does on Earth. To some people this might seem implausible, but to see the truth behind this statement we need to explore the close link between evolution and genetics.

Evolution relies upon a complex series of operations involving gigantic organic molecules such as DNA (deoxyribonucleic acid) and RNA (ribonucleic acid). These are commonly known as the 'molecules of life'. Coupled with them is a set of smaller building blocks: molecules called nucleotides that form DNA and RNA, and the proteins, enzymes and other biochemicals needed to maintain cells and sustain our existence.

But do we not have to go back even further? To begin with, how do we know that life on other planets is based upon DNA or even reliant upon the element carbon?

Well, of course at some point we have to make some assumptions. If we keep questioning at every level we hit a wall of incomprehension and can go no further. So, to avoid this we have to apply some basic principles. Scientists accept that there are certain fundamental axioms – concepts and theories that seem to be universals. The theory of relativity appears to be one of these; Darwinian evolution is another. A still more fundamental concept is what is called the 'principle of universality', or the idea that the universe is homogeneous. Put simply: 'What happens here, happens there.'

But how can we be sure this is true?

Well, one example is that we can observe distant stars from earth and determine that the chemicals present in those stars are the same as those we find on earth and present in our own star – the Sun (but in different proportions).

So where does this lead us in our exploration of the laws governing alien biology?

Firstly, we have to clear up what we mean by alien life. The chances are that there is an infinite variety of alien life forms in the universe. There may well be many ways in which life could develop and change. It would seem likely that some form of mechanism always has to allow for evolution or else the life form could not develop, but it is conceivable that this life form may not be based upon DNA. If this was the case, it is equally probable that we would not recognize such beings and would almost certainly not be able to communicate with them. Therefore, as I did in the last chapter, I would like to leave aside this possibility and restrict this chapter to an exploration of DNA-based life.

For life to have evolved in a recognizable form, it would almost certainly be based upon the element carbon, which would then allow for a biological framework involving DNA and therefore a system involving evolution via natural selection (Darwinian evolution). This is a limitation, but as we know from living on Earth (where all life forms are carbon-based) this wonderful element still offers the potential for a massive diversity of life.

So, why carbon? Thanks to the principle of universality we can safely say that carbon is the only element that will lie at the centre of a biological system yielding 'life as we know it'. The reason for this was touched on in the last chapter; carbon possesses unique properties. In many ways it is very similar to any other element, but in one vital respect it is different from any other atom in that it is the only known atom that can form the backbone of really large molecules, called *organic* molecules, and even larger conglomerates, called *biochemicals*. Also it has an almost unique ability to form long chains and rings of atoms around which other atoms can be attached. I say 'almost' because other atoms can form chains and rings, but they do not show anything like the versatility of carbon. (The best example of another element that does form chains is silicon. This element shares some of the characteristics of carbon, but because the bonds formed between silicon atoms are not so strong as those between carbon atoms, it can only form stable chains up to five or six atoms in length, and it is unable to form multiple bonds or cyclic structures, which carbon finds easy.)

Because of these facts, carbon is unique. It is the only atom able to form huge molecules, which provide the building blocks of life. And, because of the principle of universality, we know not only that this is true here in our local environment but that it is a fact of life everywhere in the observable universe. We know that there cannot be another atom like carbon, one we have inadvertently overlooked, because such an atom would not fit into what is called the periodic table – a scheme in which all the different types of atom in the universe have a strict position and interconnect in a precise pattern.

The periodic table was devised over a century ago by a Russian chemist named Dmitry Ivanovich Mendeleyev, and in it each element is allocated a position appropriate to their physical characteristics and properties. It is inconceivable that some odd element perhaps found only on the home world of the Ogrons could be squeezed in. Over the decades since it was first established, all the gaps in the periodic table have been filled, and scientists have added elements at the end of the table (unstable and very short-lived atoms found only in extreme situations such as the heart of a nuclear process), but there can be no other previously undiscovered elements that somehow fit into the middle of the scheme.

So, it is an irrefutable fact that carbon is the only element that can form molecules large enough to act as the molecules of life, DNA and RNA. Only carbon can form the smaller building blocks, the nucleotides, that go to form these massive structures, and only carbon can produce the great variety of proteins, enzymes and other biochemicals needed to maintain healthy cells and sustain our physical existence. And, because of the homogeneous nature of the universe, this is without doubt the case both 'here' and 'there'.

These biochemicals lie at the heart of genetics, and they are intimately involved in the mechanisms of evolution that are central to any discussion of exobiology. But how does this link between genetics and evolution come about?

Earlier, I defined life as a system that can reproduce and pass on mutated information from generation to generation, or as a commodity involved with evolutionary change via natural selection; and of course evolution works via reproduction.

However, this link creates what appears on the surface to be a chicken-and-egg scenario. Consider the facts: the genetic code is carried by DNA, but, if the ability to undergo evolution is a requirement of 'life', and this process itself requires an elaborate set of processes involving DNA, how did 'life' originate in the first place? To put it another way: any entity that can evolve (or by our definition be 'alive') has to be complex enough to possess the genetic material with which it can evolve. But how could an organism reach this level of complexity without evolving?

The fundamental problem comes down to this question: how could simple aminoacids that were probably around in the primeval soup on earth a little under four billion years ago have changed into what we see as 'biological material' or the simple living matter that then evolved into more advanced forms, and eventually, us?

We are not so much concerned with the second part of this question – the grand sweep from single-celled organisms to 21st-century humans. For our purposes, it is the change from non-living, or what is called 'prebiotic', material to living cells that lies at the core of the problem – the jump from unorganized RNA to a simple bacterium.

There are two theories that attempt to explain how this occurred. The first of these is the *RNA-world hypothesis*. This suggests that somehow a small quantity of a type of RNA was produced on the early earth and that this had the ability to perform a number of roles on top of the functions it demonstrates today. The RNA that is postulated would have been able to replicate without the presence of protein, perhaps by using some of the protein within its own structure, and this RNA would also be able to self-catalyse every step of the protein-production process. This may sound unlikely, but recently scientists have found molecules called 'ribozymes' which are RNA catalysts, or enzymes made from RNA. However, these molecules are still some way from RNA that could have self-replicated.

The other contesting theory to explain the jump from prebiotic to biological material is that of biologist A. Graham Cairns-Smith, of the University of Glasgow. He suggests that the organic agents that led to the formation of living things actually evolved from inorganic materials.

At first glance this too seems rather startling, because for most of us there is a vast difference between organic and inorganic materials. All living things are organic – the food we consume, the animals and plants that fill the planet. Inorganic materials include the rocks and stones, the gases that constitute our atmosphere, things generally deemed 'inanimate'.

Professor Cairns-Smith points out that, like the biochemical system using DNA and RNA, complex *in*organic systems are capable of replicating and passing on information, albeit in a much simpler way. In the system that operates in the modern biosphere, DNA carries a code in the form of an almost unimaginably complex set of instructions that acts as the blueprint for reproduction, while RNA and the proteins also play their respective roles in bringing this about. What Cairns-Smith proposes is that around four billion years ago a simpler system operated which initially did not need RNA, DNA, or even proteins.

In this system, the first step is to produce what he calls a 'low-tech set of machinery' using crystal structures present in clays. These clays, although far less complex than a DNA molecule, can create a self-replicating system in which information is passed on from one 'layer' to another. In this they mirror the way DNA replicates.

From this low-tech start, Cairns-Smith supposes that a gradually more complex system evolved which incorporated organic molecules. These unsophisticated systems developed over time into the 'high-tech machinery' we have today in which DNA, RNA and proteins facilitate genetics and allow the evolution of living things via natural selection.

So, this then leads us some way towards a model for how life may have begun on earth, and it is a theory that could probably equally well explain an evolutionary mechanism for any DNA-based life form existing on any number of worlds throughout the universe. But given these possible mechanisms for how life began, how would life on other worlds have developed from this simple start? Could we expect biological processes to have followed a similar route to the way they operated here? And, if so, would it lead to human-type beings or very different creatures?

To find answers we must turn to the discipline of 'developmental biology'. This is really a blend of several linked fields, including evolutionary biology, palaeontology and genetics, which each help developmental biologists understand how creatures on earth have evolved from simple forms dating back billions of years, to the fauna and flora we have on earth today.

They start with two other disciplines from biology. Firstly, they need to consider 'genetic retracing', which involves analysing how genetic material has changed over long periods. Genetic characteristics are of course one of the key factors in determining the nature and diversity of life, and it is possible to trace the way genetic material has altered both within species and across different species. Biologists can then reach conclusions about common ancestors of modern species, and this gives them the information to construct a 'family tree' dating back many hundreds of millions of years.

The other approach is 'evolutionary biology', which looks at the range of modern animal structures, or the 'body plans' of animals – the fundamental groupings of different animal types we see all around us. From these it is possible to work backwards using computer models to determine the original plans.

Neither of these techniques is straightforward. Evolutionary biologists use powerful computer models that utilise a vast collection of parameters. These are primarily based on information gained from palaeontological and archaeological finds and they use these parameters to determine how 'function' links with 'design' and how organisms have adapted to their environment. Genetic retracing is a complex science, because genes (like species) evolve at different rates and the evolutionary lines can branch in elaborate ways.

Life probably first appeared on earth around 3.85 billion years ago, and although we are not sure how the change from prebiotic material to living organisms occurred, it is clear that once it did, life flourished and evolved on this planet. But it was certainly no simple development.

Until about 550 million years ago, the most complex form of life on earth was an organism no more advanced than simple algae. Even then, algae were new arrivals, first appearing about one billion years ago. For the 2.85 billion years leading up to this point, life consisted of nothing more elaborate than single-celled organisms such as bacteria.

Then, during what is called the 'Neoproterozoic era', more advanced organisms began to appear. These simple creatures probably resembled modern-day sea pens, jellyfish, primitive worms and slug-like animals, all of which have left faint fossil remains and markings.

But then suddenly, around 530 million years ago, there was a complete and, in a geological sense, sudden change in the evolutionary development of life on earth. Throughout the Neoproterozoic period, the earth had been populated by animals which displayed a relatively small collection of different body plans, but then everything changed. Within a short space of time the earth was populated by a vast range of different creatures.

This transformation is called the 'Cambrian explosion', a burst of activity in the life of the fauna of this planet. Before it, the Earth was home to only relatively few simple organisms; after it, Earth was populated with organisms that, although still simple, evolved into almost every known type of shelled invertebrates (clams, snails and arthropods). In time these gave rise to modern vertebrates, then mammals, and eventually humankind.

Most importantly for exobiologists, after the Cambrian explosion the basic body plans of all animals on earth had been established. From that point on, all evolutionary steps (including what some consider one of the most dramatic – the point at which some animals left the sea to live on land) merely required subtle refinements of the basic animal types established during the Cambrian explosion.

Amazingly, this single dramatic change in the history of life on earth produced just thirty-seven distinct body plans, which account for absolutely every animal on the planet today. Furthermore, it is now known that almost all living things share a collection of genes called 'regulatory genes' that determine the essential body plan of a creature.

Most animals start from a single cell, the fertilized egg or zygote. This cell then divides and multiplies and specialized parts are formed – the organs, glands, skin, bone, muscle tissues. But every organism has a set of common genes that control the process of protein formation that leads to the formation of these parts. As this process continues, the genes required become more and more specialized.

The simplest genetic commands are those that determine the body axes of an embryo – which end becomes the head and which the tail, which is the back and which the front. These instructions are the most basic and are examples of a shared characteristic between almost all species. Further along the process, a collection of genes determine whether a head is developed outside the trunk of the body (as in most animals), while others instruct the growth of limbs.

In species as different as, say, bats and eels, these instructions will be very different, but for a sheep and a dog, or even a sheep and a human, they will be much more alike. It is only when we consider what are comparatively specialized functions and characteristics of a creature that we see very clear differentiation. Even the bat's wings and the forelimbs of a sheep can be thought of as controlled by a similar set of regulatory genes.

The origins of this process are ancient. The DNA sequence – a vast and complex set of instructions controlling the process – was certainly operating on earth during the Precambrian period, before the Cambrian explosion 530 million years ago. Indeed, it had to be in place to allow the explosion to happen.

The DNA that does this is found in a collection of regulatory genes called 'Hox genes' that are themselves clustered in animal chromosomes. Amazingly, these 'Hox clusters' may be thought of as 'templates' for the animal of which they are a part; the genes are literally arranged in the gene cluster in the precise way the animal parts they control are positioned in the growing embryo. So, genes that control the development of the head are at one end of the cluster, the wings or legs are part way down, and the rear end of the cluster holds the genes that control development of the lower parts or rear of the animal.

So, what does all this mean for the exobiologist?

Firstly it shows that only a few dozen patterns or layouts (body plans) are needed for a vast array of different species. Secondly, if we restrict our vision of alien life to those based upon DNA, we still end up with a universe offering an incredible variety of shapes and forms.

And what does this mean for the chances of there existing on other distant worlds a massive array of monsters, the malevolent aliens, humanoids and reptoids and others who regularly clash with *Doctor Who* and many other science-fiction heroes?

A way to answer this may lie with the currently controversial idea of 'convergence', the essence of which is the proposition that 'Many very different starting points produce a limited number of solutions to a task.'

An everyday example is an aircraft. The 'task' is to build an efficient flying vehicle for a reasonable price that will transport a small group of human beings from A to B at high speed, safely and comfortably.

Now, before aircraft were invented people may have thought that there was a vast range of ways of doing this. Indeed, the experiments to make the first planes were extremely varied, and today, a century after the first heavier-than-air machines were flown, there is an array of different flying vehicles. But these fall into a limited number of types (analogous to body plans in the animal kingdom).

In fact, differences between aircraft are all rather superficial and are largely restricted to appearance and size, and subtle refinements in style, layout of the interior and alterations to fit specialized tasks. At their root, they are all metal cylinders with doors, wheels, wings, engines and tails, they all use fossil fuels, they have a front and a back and seats on which humans sit and they all travel through the skies (a helicopter is a little different, but even this has many of the same characteristics).

In the same way in which humans have solved this particular task, Nature deals in a limited number of different ways with the design problems it faces. An example is the eye. The eyes of all advanced animals are the same, they have a lens, an iris, a retina, optic nerves. Simpler creatures have less sophisticated eyes, but those of, say, mammals have followed a similar avenue of evolution to end up with a product that is almost identical. Most crucially, Nature always performs this task in the most efficient way possible. Scientists refer to this process of development as following a path of 'perfectibility'. And it works excellently well here, so it should work soundly on other inhabited worlds.

So, what limitations are there to the types of DNA-based life forms we could hope to find one day on another world?

To establish any kind of informed answer we must look at two distinct factors. We have to consider first the evolutionary factors on an alien world, and second, the environmental conditions.

Key to the evolution of life on earth were the great events of the Cambrian explosion, and it is only by looking at this singular example that we can formulate an idea of what could have, and might still be happening, on other worlds. There are two competing theories of what precipitated the events on earth 530 million years ago. The first is simply that the time was right, that Nature had experimented with various evolutionary mechanisms and eventually hit upon the right way forward. In saying this, I certainly do not want to imply that Nature in any way 'planned' this or 'knew' what to do. Nor is there anything to say that it was in some way 'guided' by an external agent, be it God or an alien in a shiny spacecraft deliberately setting in motion a process that would lead to the production of a dominant species like Homo sapiens. Natural selection does not operate by any sort of plan: it is driven by success and buffeted by random events, needing neither a God nor an alien intelligence. Instead, in one sense, this explanation for the Cambrian explosion is linked to convergence – Nature had simply found a path from point A to point B by trial and error.

If this is the case then we can feel confident that a very similar process will have occurred on any number of alien worlds. It would be a fundamental and very simple process, a consequence of life reaching a certain level of complexity, a point from where it is spurred on to the next stage.

The rival theory, or, rather, group of theories, offers a less rosy picture for the chances of there being advanced life in abundance, because it suggests that the Cambrian explosion was precipitated by some unknown freak ecological event. Perhaps there was a giant meteorite or asteroid collision. Another alternative is the possibility that something caused a massive change in atmospheric conditions on the planet. An increase in the percentage of oxygen in the atmosphere would have almost certainly triggered a dramatic surge in biological activity on the surface and this could explain why a vast array of fresh life forms appeared on earth within such a brief period.

At present, nobody can say for sure which of these theories is correct, but the answer will have great bearing upon our vision of the type of universe in which we live. One theory claims to offer the key to a highly populated universe, the other a bleaker (but not totally lifeless) alternative.

One argument in favour of extraterrestrial life is to say: it happened here, and, with what are certainly a large number of planets to choose from, it could happen elsewhere. After the initial spark of evolution from the Cambrian explosion there is a long, hard road of development that leads to any form of advanced life, one capable of developing a civilization. But if for our purposes we assume the

Cambrian explosion was inevitable and something similar has happened elsewhere, can we then say anything about evolutionary routes on other planets?

Once again, we can only make comparisons with the one example we have – the evolution of life on Earth. To arrive at an answer to the question of whether or not intelligent life could have evolved on other worlds we have to look at what 'intelligent' means as well as its relevance to the ability to form a civilization.

As I mentioned in the last chapter, on Earth, the only animal with any form of conscious social interaction or intellect (as distinct from pure intelligence) is Homo sapiens. We are the only animals on the planet to keep records, to have developed a recordable language (a form of writing), to have built a civilization based upon trade, and, crucially, to plan, and to have a concept of our place in the world and the flow of generations of our species.

So perhaps the first question to ask is: what is it about us that makes us different from the other species on our own planet? If we can arrive at an answer to that, we might be able to extend the principle to extraterrestrials.

The difference between us and other species seems to come down, at least in part, to brain size. We have very large brains for our bodies. If the human brain was unfolded and spread out, it would cover four sheets of A4 paper. By comparison, a rat's brain would barely cover a postage stamp. However, it is not just a matter of size, but the way the brain is used. Dolphins have very large brains, but it is believed that most of their brain capacity is involved with managing their complex sonar system.

The reason brain size is so important for humans is that it relates to our development of incredibly complex language skills. And language is a basic (but not the only) requirement for civilization and social development.

In the pre-history of the human race, we know there was a 'sudden' four-fold increase in brain size between 1.5 and 2.5 million years ago. It is thought that before this point, our early human ancestors had a brain capacity comparable to that of a chimp (about one quarter the size of the modern human brain). What caused this rapid development remains a mystery. The most likely explanation is that our ancestors were faced with a severe 'challenge' to their continued existence, and they only survived by developing larger brains over a period of many generations. Whatever the mechanism for this change may have been, it marks another key turning point in the development of the human race, another crucial jump along the evolutionary road.

The most likely thing that presented early humans with a challenge that was both global and sufficiently powerful was the arrival of the most recent Ice Age, the 'Quaternary Ice Age', which is believed to have acted as a 'survival filter' for many species, including the species that evolved into Homo sapiens.

This survival filter is really a description of the mechanism via which some species overcome problems, develop skills and survive, while others fail and fall. Clearly, early Homo erectus (the immediate progenitor of the first Homo sapiens) learned a great deal from the experience of the Quaternary Ice Age. One of these new skills was the ability to meet dietary needs by being less choosy than other species.

By observing how Homo erectus survived and evolved, biologists have concluded that the more intelligent land-based animals are always omnivorous. This, they think, is because omnivores can adapt their tastes to find diverse sources of food, and the effort to search out new resources is also a learning process which helps the animal develop skills not common in carnivorous or vegetarian animals. In the case of Homo erectus, those who survived the change in the environment did so by learning to find and use new resources, which led them to gradually develop social skills, to create communities, to develop language and to eventually take the first steps towards civilization.

And with language comes what we call 'intellect'. If we define intellect as the ability to communicate and process ideas, then a huge leap in human evolution came about with the development of syntax, and from that the ability to string together meaningless sounds (phonemes) to make 'meaningful' words. This ability enables us to create sentences and to communicate abstract ideas, to plan, to create social rules, taboos and hierarchies. Language really is the cornerstone of civilization.

But would there necessarily have been events comparable to the Ice Ages on other worlds?

It would seem very likely. Although there are a number of plausible theories to choose from, nobody knows for sure why the series of Ice Ages occurred on earth and whether these are linked to very common natural processes in the life of a planet. But it would seem reasonable to suppose they would be common on a good proportion of earth-like worlds. And, of course, Ice Ages may not be the only form of challenge an embryonic dominant species might be offered. Other worlds may suffer environmental changes precipitated by comet or asteroid collisions, volcanic activity or short-lived irregularities in the behaviour of the planet's sun.

But what other evolutionary considerations should be taken into account? Brain capacity and brain application are only part of the developmental formula, albeit crucial ones. Another extremely important factor is physical versatility. And this leads to the question: could only humans have created a civilization on earth? Why, for example, have dolphins not achieved the same status?

Dolphins are highly intelligent creatures but they exhibit a form of intelligence that appears to be very different from our own, one which has not led them to develop intellect, to create what we understand as a 'civilization' or a 'society'. But, why is this?

The simple truth seems to be that dolphins did not have a chance of competing with humans because they live in an environment that makes it very difficult for an intelligent animal to create any form of infrastructure. This is due to several complex factors.

Dolphins do not have digits with which they can manipulate materials, and they certainly have not developed opposable thumbs which have been one of the most important distinctions between human and non-human primate development on earth. Dolphins are a very successful species – their physiology has evolved in a way that allows the animal to be perfectly adapted to its environment – but they have been unable even to start on the road to civilization.

Dolphin 'language', although sophisticated compared to those of almost all other animals on the planet, has not evolved in a way that can lead to social development past a rudimentary level. With the bodies they possess, they could not have constructed the aquatic equivalent of buildings; they cannot easily record any knowledge they acquire in the way humans do, using writing; they could not easily cultivate their territory or manage other animals, which means they are constantly at the whim of fluctuations in food supply. Finally, they could not have developed weapons, so they could not have engaged in wars, which are, like it or not, extremely important to the development of any technological society. So, extending this example, it might be fair to say that any intelligent aquatic animal has only a slim chance of developing a civilization as we understand the concept.

This then means that any planet which brings forth life must have enough land to allow animals to develop and for the correct ratio of plants and animals to arise in order to create a balanced ecosystem. Furthermore, any world which is entirely covered in water will almost certainly not harbour life any more advanced than the equivalent to Earth-style fish and sea mammals. These animals would be facing an almost impossible struggle to form an organized society or advanced civilization.

Turning to the other strand of the arguments to determine the nature of alien life, what special environmental conditions aid the development of civilization?

Bacteria are known to survive in some extreme environments on Earth – in radioactive waste, thousands of feet beneath the seabed, in hot springs and in the frozen wastes of Antarctica. They are of course extremely hardy creatures; more sophisticated animals could not have evolved within environments as harsh as those in which bacteria appear to thrive. This is an example of a fundamental rule of biology, that the simpler the organism the hardier it is.

This places limits upon the development of complex life forms. Firstly, the temperature of the environment must lie somewhere in the region of 0 °C (273 K) to 40 °C (313 K). The reason for this is that enzymes that are essential to biochemical processes cannot operate at temperatures much higher than 40 °C, at which point they begin to denature. By the same token, because all biochemical processes require water as a medium, they cannot occur below 0 °C (at a pressure of 1 atmosphere), which is of course the freezing point of water. A caveat to this is that on a planet with a different atmospheric pressure, water will freeze at a different temperature, but, as we will see, environments with atmospheric pressures very different from that on Earth present their own problems for the evolution of life as we know it.

One further environmental factor of great importance is the level and type of radiation present on an alien world. For life to flourish, the environment must not be flooded with intense radiation, as this damages biochemicals and inhibits many of the chemical reactions required for mechanisms that control the functioning and continued growth of cells.

All these restrictions indicate that the environment of any planet where complex multi-cellular beings have evolved must fit into a relatively narrow set of types. On the one hand, it must provide the right chemical conditions for life; but on the other it has to offer a challenge to living things so that natural selection can operate and evolution can occur.

If we consider first the atmosphere of a candidate world, what are the limitations that must be in place to encourage life? Would it be possible to have a successful ecosystem on another planet that supports DNA-based life but does not have an atmosphere similar to that of the Earth?

The answer is almost certainly 'no'. The reasons for this are complex. On Earth we live in a balanced ecosystem in which plants need carbon dioxide to facilitate photosynthesis, which in turn produces oxygen. All animals use oxygen, which is absorbed by the blood and carried to the cells of the body, where it is involved in almost all of the biochemical mechanisms that maintain our bodies. Furthermore, no alien creature could evolve or remain alive on a planet without interacting with other organisms. All organisms must be part of an ecosystem, and any ecosystem must include gaseous cycles similar to the oxygen–carbon dioxide cycle on earth. Such alien systems would have to integrate organisms similar to our plants and animals.

An alien world may have an ecosystem based upon animals and some other kingdom – perhaps some form of animate mineral or living rock – but the same situation applies to this as it does to the animal/plant relationship on earth.

This does allow scope for an atmosphere with *proportions* of the gases different from the ones we have in our atmosphere. But, even then, it precludes excessively high concentrations of any gases that

DNA-based life would find toxic. It is known that only some very rare bacteria can survive in an atmosphere in which there is very little oxygen or a large amount of a toxic gas, such as methane.

What of other environmental considerations? What limits must be placed on atmospheric pressure or gravitational fields? How would these affect the diversity of life on an alien world?

On a planet where the atmospheric pressure is higher than on Earth, it is possible that intelligent creatures could have evolved that look very different from humans. The layout of the respiratory systems of such creatures would probably be very different, because the pressure of the gases they breathe would not be the same as it is here, which means that the processes allowing gases to diffuse into their version of a circulatory system would operate at different potentials. But this certainly does not exclude the possibility of such creatures reaching a high level of development.

More difficult to resolve is the effect of high or low gravity. The strength of the gravitational field on an alien world will greatly affect the body types and the behaviour of creatures living there.

All land-based mammals on earth fall within a relatively narrow range of size – there are no mammals 200 feet long or insect-sized. But if the force of gravity was, say, thirty times greater than it is here, the creatures that developed on such a world would be very much smaller than those we have on Earth and they would be far less mobile. In extreme conditions we could imagine intelligent creatures that were almost flat.

Conversely, planets with low gravity would bring forth creatures that were much larger but lighter, and they would probably move by natural methods of flight rather than moving around on the surface as we do.

However, a great difference in physical size from that we witness on Earth creates its own problems. Think back to the aircraft analogy for a moment. At either end of the scale, there is a definite limit to the size of aircraft that could be of practical use to humans. In the same way, very large creatures present design difficulties for Nature.

One of the major problems for very large animals (and there are many such difficulties) is the fact that they would require huge hearts to supply the volume of blood needed to keep their bodies functioning, and big hearts need big lungs. As we know from the design of animals on Earth, this is not a strict limitation, but for an animal to evolve into the dominant species and to establish a civilization, they also need a large brain. Such a brain would need a large head and still more blood to supply the cells with oxygen, which needs a massive heart and lungs: and so we enter a vicious circle.

Some claim that the 'success' of the dinosaur refutes this argument. It does not. Using this argument, 'success' is equated simply with longevity. The dinosaurs were around for many millions of years, but in any definition of a successful species that evolves to create a civilization we have to take into account the role an animal plays in the ecosystem of which it is a part. Humans are the only animals we know have created a civilization, the only species that controls its environment in any large-scale way.

So, even if we only consider DNA-based life it is clear that there could be a range of different shapes and sizes for extraterrestrials, but these have to fit into certain limits; and it may be that on planets with approximately the same gravity as the Earth, a species that formed a civilization there would be at least vaguely similar to humans.

But what of the details, what exobiologists call 'parochial characteristics', or 'cosmetic differences'? What about the number of limbs, sensory apparatus or colouring an alien may possess? Is it likely that we will one day encounter a two-headed, five-legged green blob offering the tentacle of friendship?

Whether or not an extra pair of limbs or a third eye would be favoured within an ecosystem on another world is open to debate. There may be advantages in having these things. But as we saw earlier, within any environment, Nature will always go for the most efficient option. If the advantages of a third eye or an extra pair of ears outweigh the demands produced by the extra weight, the extra blood requirements, the development time (both in terms of evolution and within the womb), then it could happen. If not, then evolution is unlikely to allow such creatures to dominate, and they would easily be made obsolete by better models.

But, how far can we go with this argument?

There are those who do take the anthropomorphic argument to its limit and suggest that the most likely design for any successful life form that has evolved to the point of developing a civilization will be similar to ours – that they will look like us (or us like them). But why?

Consider the number of limbs a creature has. Do we need more than two legs? It is possible that a creature developing on a world with high gravity would need three legs or more to allow it to move more efficiently under the strain, but then it could be argued that a biped could have evolved on the same world that has two much better legs, allowing it to move around more readily than its multiped rival.

Does any creature need two heads? We have two of almost everything else, why only one head?

Quite simply, two brains would require too much blood for the same sized heart, and we are back in the same cul-de-sac as we were in with the 'size of lungs versus size of animal' loop. But, there is also the fact that convergence would not allow for such a consequence. It would enforce the most efficient solution – a biped with one head is better than a large ungainly biped with two.

So, a three-legged, two-headed beast may never get through evolutionary quality control. Indeed, it may be argued that this latter body type is unlikely because among the great diversity of life on Earth there has never been a two-headed anything. Of course, strictly speaking, this is not an empirically sound argument, because we are trying to deduce outcomes for an alien world where (within the limits required for DNA-based life) conditions could be quite different.

What may we conclude from these arguments? Firstly, most scientists believe that life in the universe is plentiful, and many would put money on the idea that our civilization is just one of many. Less convincing is the argument that there could be intelligent life based upon anything other than DNA-led biochemistry. If there are creatures that have flourished and become highly evolved via such a route, we may never encounter them and would almost certainly never be able to communicate with them.

If we consider the narrower arena of life based upon DNA, we still have a truly mind-boggling range of possibilities; just look at the marvellous variety of life on Earth. DNA-based life, it is argued, will have formed and evolved on a range of planet types, but there are certain restraints, certain environmental limits, that must be respected. These, along with caveats relating to the way evolution could progress on alien worlds, narrows the range of life forms somewhat and can lead us to believe that the most likely shape for an intelligent, DNA-based alien approximates to the humanoid form. But variations in the details could produce very different creatures.

It is possible that an alien species could use different chemicals to carry oxygen to their cells, and their sun might emit a slightly different range of radiation. Combined, this means that such aliens would be a different colour from any human, because it is the colour of the haemoglobin in our blood and the amount of melanin in our skin that determines our colour.

For the creators of science fiction, the only limitation is of course the range of their imaginations. What this scientific survey shows is that other worlds could contain a vast multitude of extraordinary life forms of all shapes and sizes, colours and textures, just as there is on Earth, where millions of species co-exist. However, if we are to limit our scope to the nature of intelligent life within a realistic universe then, yes, little green men are possible, but they could not be too little; giant bipeds with tentacles for arms are allowed, but these giants could not be too big, and intelligent amoeba are definitely out. Yet, of course, this still leaves the possibility of massive diversity. Just look around and consider the fantastic range of life on Earth and, indeed, the wonderful diversity of form within just our own species. If there are millions of life-bearing planets out there, each with its own subtly different environment, life in the universe must indeed be rich and more varied than even science fiction writers could possibly imagine.

4

Gallifrey or Bust

Is Interstellar Travel Possible?

How often have I said to you that when you have eliminated
the impossible, whatever remains, however improbable,
must be the truth? Sherlock Holmes

Almost every episode of *Doctor Who* is set on Earth or on a planet somewhere or somewhen with very little reference made to its location. Only rarely are spaceships shown, and the plots usually have nothing to do with the way the aliens arrived on Earth; and if the Doctor is a visitor somewhere he simply arrives there in the Tardis, which is able to travel through both space and time with equal ease.

In other science-fiction stories, the means by which aliens travel around the universe is sometimes covered in more detail. Arthur C. Clarke and Isaac Asimov wrote wonderful stories during the 1940s and 1950s in which, using their scientific knowledge, they speculated on how interstellar travel might be achieved. For TV series *Star Trek* or *Babylon 5* and films such as *Star Wars*, interstellar travel is a central strand of the story, and the ability of advanced extraterrestrials to travel quickly between planets and star systems is a given. But how realistic is the notion of rapid travel between solar systems?

It would be necessary for any advanced alien culture to have mastered interstellar travel for them to be visiting the Earth, as the UFO enthusiasts believe they do; and if humans are to ever explore planets beyond our own backyard we too, would have to develop the technology to travel extremely fast, to somehow find a way of getting round the limitation of the speed limit to the universe – the speed of light 'c'.

The facts of interstellar travel all revolve around distance, time and power. Because the distances between the stars are unimaginably huge, the time needed to travel interstellar distances is correspondingly large, and any system which may have a chance of overcoming this restriction requires huge amounts of power.

The problem begins with Einstein's special theory of relativity, which I described in Chapter 1. This shows that if *c* (the speed of light) is constant, space and time must be relative. The consequences of this are that the pilot of a spaceship A, observing a moving spaceship B, will see the light arriving from the ship as travelling at velocity *c* irrespective of his own velocity or the velocity of spaceship B. This means that time will be measured differently on spaceships A and B. As A travels faster, time according to the pilot of A will slow, any measurement of linear dimensions will also change relative to B, and the mass of A will increase. If spaceship A could travel *at* the speed of light it would have infinite mass, it would shrink to nothing, and, on board, time would stop completely.

As I said in Chapter 1, this is not a concocted story designed merely to spoil our fun. These are universal, irrefutable facts, and the theory of relativity has been proved correct by a century of experiments. Indeed, it is rather ridiculous to still call the theory of relativity a theory at all.

Now, you might wonder, if the consequences of relativity are true facts, then why don't we experience them every day of our lives? When we are travelling along the motorway why don't we grow heavier? Why doesn't time slow down and why is a metre still a metre? Well the truth of the matter is that we *do* experience the effects of relativity when we drive along the motorway, it is simply that at the speed we are travelling (say 120 kph) we do not notice the changes because they are so incredibly small.

A recent shuttle mission showed how minuscule are the effects of relativity at low speeds. Travelling in orbit at a sprightly 8 kilometres per second, clocks aboard the shuttle ticked less than one ten-millionth of a second slower than their counterparts on Earth. At CERN, the giant particle accelerator near Geneva in Switzerland, sub-atomic particles are routinely accelerated to near-light speeds, and their masses are seen to increase precisely as Einstein's calculations predict.

The law that states that no material object can travel at the speed of light is an undeniable, unbreakable one. Consequently, the only possible ways in which a very advanced civilization might cross interstellar distances are either by travelling at speeds that are not impinged upon significantly by Einstein's theory, but which get them to their destination eventually, or by working out some novel way around that theory. But before considering the possible ways we could cheat on Einstein, let's take a look at some less ambitious ways of travelling very fast.

In his story *From the Earth to the Moon* (1865), Jules Verne described a method of sending a rocket to the moon that involved firing the vehicle from a huge cannon. Since then, both scientists and science-fiction writers have come up with a range of ingenious propulsion systems to facilitate interstellar travel. These include fusion drives, antimatter engines, spaceships that utilize the weird properties of wormholes, and space-warping devices such as those used by the starship *Enterprise*.

All conventional space propulsion systems (and by that I mean engines that do not use some exotic property of space itself such as warping or wormholes) must work on the principle of Newton's third law of motion, which states that, 'For every action there is an equal and opposite reaction.' In this way, a spaceship is no different from a jet aircraft – material is expelled from the back of the craft and the craft moves forward; simple. The limitation for the hopeful interstellar traveller is how quickly the vehicle can move forward.

The spacecraft we have developed so far all work by using the power generated by chemical reactions. The greatest energy requirement has been that needed to escape the Earth's gravitational pull – to achieve an 'escape velocity' – so that the Saturn V, a shuttle or the Ariane craft can carry their payloads into orbit. On Earth, the escape velocity is about 25,000 kph. All manoeuvres aboard the Apollo craft during their journeys to the Moon depended upon relatively small engines and thrusters that expelled hot gases from their exhausts and adjusted the course of the spaceship. Without these, the capsules would have been entirely at the whim of the gravitational forces at work beyond the Earth's atmosphere.

The next level of sophistication is some form of fission-powered spacecraft engine. This is the power source used in nuclear reactors and unleashed in the earliest atomic bombs. When large, unstable atomic nuclei are made to decay, or undergo fission, they produce energy. The value of this energy depends upon the mass of material undergoing fission and can be calculated using Einstein's equation $E = mc^2$.

Although this is the most powerful controllable energy source we have currently, it doesn't come close to providing the energy needed for a vehicle to reach the stars, and the mass of fissionable material required even for efficient interplanetary travel (within our own tiny solar system) would be so large there would be little room left for crew or cargo. A more powerful form of nuclear energy comes from a process called nuclear fusion.

In 1989 there was a brief flurry of excitement when two scientists, Martin Fleischmann and Stanley Pons, claimed they had devised a technique called 'cold fusion', which appeared to require nothing more than a pair of electrodes and some commonplace chemicals placed in a jar. Sadly, the excitement died when the experiments proved unrepeatable and the hopes of scientists returned to conventional fusion. This is the mechanism by which our sun or any other star is powered. In the laboratory the process involves fusing together small nuclei such as deuterium and tritium (which are heavy isotopes of hydrogen, i.e. ones that contain more neutrons than the most common form of hydrogen) to produce large amounts of energy.

For almost sixty years scientists have been trying to develop practical nuclear fusion because it has many advantages over fission. It is a far more powerful nuclear process, and it is relatively clean because it does not use dangerously radioactive elements such as uranium-238, which is converted into plutonium-239 in modern fast-breeder fission reactors. These isotopes remain dangerous for hundreds of thousands of years.

These are the plus points of fusion power. The downside has so far been the problems of containment and efficiency. In order to bring about fusion, temperatures of around 10 million °C are required, the sort of temperature produced at the sun's core. Under such extreme conditions the positively charged nuclei of the hydrogen isotopes are forced to overcome their electrostatic repulsion. But this fused material exists as a superheated plasma that cannot be kept in any form of physical container. Furthermore, current methods of bringing about fusion in the lab show negative efficiency; that is, the energy put in far exceeds the energy return.

That said, scientists hope to crack both these problems in the future, and fusion energy is seen as the most likely way in which we could overcome the Earth's looming resource crisis, and power space vehicles to the planets of our Solar System.

Assuming another civilization is only a few decades ahead of us, they would almost certainly have developed fusion power, and if they are further advanced still they would have mastered the use of fusion engines aboard spacecraft. Unfortunately, though, even this energy source would be of little practical help in constructing a vehicle that could travel to the stars. The simple reason for this is that in order to accelerate the craft to even a small fraction of the speed of light so much fusible material would be needed that such a craft would be quite impractical.

It has been calculated that to accelerate a fusion-powered spaceship to just 10 per cent of the speed of light would require about 15 times its mass in fuel. And this is to accelerate just once. If the craft wanted to stop at its destination it would need to use more fuel, equivalent to 15 times the mass of the ship as it then was. If we assume the outward voyage has used up half the fuel, a further 7.5 times the mass would be needed to stop. So, one start and one stop would need 22.5 times the mass of the craft. A return trip would require 28.125 times the original mass of the vessel.

A variation that avoids the need to carry such large amounts of fuel is the 'fusion ramjet', which would effectively pick up fuel as it travels. Interstellar space is not a complete vacuum: it contains hydrogen atoms, albeit ones that are distributed very finely between the stars and planets. A spacecraft could be designed with large scoops drawing in the hydrogen atoms to use as fusible material. The objection to this has always been that there is insufficient material available in space, but if the craft is moving quickly enough it would behave like a giant sea mammal drawing in plankton, or like a person running through light rain getting soaked because they are meeting the raindrops as they go.

In the future, when we first send people to the planets of our own Solar System, we will almost certainly use some form of fusion power to do this. It is a practical system for interplanetary travel, as acceleration to speeds of around 100,000 kph would be relatively easy, allowing us to get to Mars in about three weeks. But, we must remember that there is no comparison between interplanetary and interstellar distances. A spaceship using fusion power that could achieve speeds of 100,000 kph would still take a thousand generations merely to reach the nearest star, Proxima Centauri.

Putting aside fusion power, there have been a number of other suggestions for ways to achieve a reasonable fraction of light speed using conventional physics. One such idea is to use the power of nuclear explosions to thrust the craft forward.

Designers of a theoretical vehicle known as *Orion* visualize using a store of thermonuclear warheads individually propelled from the back of the craft at the rate of one every three seconds. The hot plasma produced by the explosions would impact on a 'pusher plate' propelling the spaceship forward. Unfortunately, to achieve a speed of just 3 per cent of the speed of light would need almost 300,000 one-megaton bombs.

A variant on this was *Project Daedalus*, investigated by the British Interplanetary Society during the 1970s. This theoretical system involved a craft similar to *Orion* but powered by 250 nuclear explosions per second. It was calculated that the vehicle could achieve some 12 per cent of light speed, or 130 million kph, but you can imagine why it remained nothing more than a theoretical design and was never taken very seriously.

More promising than any of these schemes is the possibility of using exotic material known as 'antimatter'.

All matter in our universe is made of atoms. These in turn are composed of what are called sub-atomic particles – neutrons and protons (which exist together in the nucleus of the atom), along with electrons which surround the nucleus. This much was understood by the beginning of the 20th century. But with the advent of quantum theory (discussed in Chapter 1), in 1929 the physicist Paul Dirac was able to use some pretty innovative mathematics to predict that all the known sub-atomic particles could have counterparts with opposite properties. These opposites became known as 'antiparticles'.

Protons are positively charged. An antiproton would have the same mass and exist in the nuclei of 'anti-atoms', but it would be negatively charged. An 'anti-electron', or 'positron', as it has become known, would be positively charged and, like the electrons surrounding the nucleus of an atom, it would exist outside the nucleus of an anti-atom. The crucial property of these antiparticles that may help us produce a powerful means of propulsion is that when matter and antimatter come into contact they annihilate each other instantly and produce energy.

In Paul Dirac's day antimatter was merely a theoretical construct, something that had popped out of the equations when he had combined the mathematics of quantum mechanics, electromagnetism and relativity. At that time, the existence of antimatter could not be demonstrated. Antimatter is not found naturally in our universe because it disappears immediately it comes into contact with matter. Today, though, we can manufacture very small quantities of antimatter in a particle accelerator.

To make an antiproton, 'normal' protons are sent whirling around the accelerator ring of a cyclotron such as the one used at CERN or at Fermilab near Chicago. In the cyclotron they are accelerated inside an intense magnetic field until they reach about half the speed of light. They are then allowed to collide with the nuclei of metal atoms. Along with X-rays and various forms of energy, this collision produces short-lived pairs of particles and antiparticles. The antiprotons are then separated from the protons before they can interact and annihilate each other.

To use antimatter as a propellant we need to allow a controlled annihilation of particles and antiparticles so that the heat energy produced can be used to drive our spacecraft. A simple design for just such a system is already on the drawing board. The idea is to fire a tiny quantity of antimatter into a hollow tungsten block which is filled with hydrogen. The particles are instantly annihilated and the energy released heats up the tungsten block. Cold hydrogen is then squirted into the centre of the device, where it is rapidly heated to about 3000 K and fired out of the engine.

The great advantage that antimatter drives offer is that they need little fuel to produce an effective acceleration. The great disadvantage is the difficulty involved in producing anything like usable amounts of the stuff. Currently, only one sixty-millionth of the energy used in producing antimatter in the world's particle accelerators ends up as particles, which is one of the reasons its current market value is about $10,000,000,000,000,000 per gram.

If we wish to use antimatter as an energy source we must also find practical ways to handle and manipulate it. The only way to manage antimatter is to use special magnetic containment systems which prevent antimatter coming into contact with conventional matter; this requires the use of extremely powerful electromagnets that consume huge amounts of energy.

None of these difficulties precludes the use of antimatter by advanced civilizations. Just because it is extremely difficult for us to use this power source today does not mean a slightly more advanced civilization would be unable to employ it. If we consider our own history, it is clear that, given the right motivations, tremendously rapid development from concept to practical application can take place. An example is the development of nuclear weapons during the first half of the 20th century. It was only in 1919 that Ernest Rutherford discovered that the nuclei of certain atoms could be made to disintegrate by bombardment. Within just twenty-six years this discovery had led to Hiroshima and Nagasaki.

Current technology may mean that antimatter is prohibitively expensive to produce, but within two or three decades this may no longer be the case. And, as we have seen, such time spans are relatively meaningless when we consider the possibility of advanced societies developing on other worlds.

The option of antimatter propulsion systems offers hope that interstellar travel may be a possibility, but even if an advanced civilization has realized the full potential of this technology, they would not be able to circumvent the natural laws of the universe and would remain limited to sub-light speeds. But then even at speeds close to light speed, journey times remain prohibitive. Imagine a journey of 50 light years from the alien's home world to Earth. At 0.95 c (95 per cent of the speed of light), this would take 52.5 years to complete, one way. 52.5 years to the people back home, that is. But because of the consequences of special relativity, as we travel faster, relative time slows, so to the crew of the spaceship this 52.5 years will be just over 15 years. But this is still far too long to be of practical use. Even if we assumed that typically aliens lived longer than us, one and a half decades is still a long time to remain on board a spacecraft and such a vehicle would be very slow for the purposes of colonizing or visiting many worlds. A way round this problem might be to use a form of suspended animation or even cryogenics, but there remain other difficulties with this scheme.

Crews sent out on round-trips of over a century might return to their home worlds to find the political structure changed. The organization that sent them may no longer exist. Such a crew would find all their relatives either dead or ancient, and almost everything once familiar would have changed enormously. Imagine a human explorer setting out on such a mission in the year 1901 and returning to Earth in 2006. They may have aged just over thirty years, but the world they left would be almost unrecognizable to them.

If we imagine, as UFO enthusiasts do, that interstellar travel is commonplace and that aliens visiting us operate as part of an organized federation or interplanetary authority, we must accept that they could not travel at sub-light speeds. First, any form of command structure would be impossible to maintain over such distances and timescales. Second, it is surely a universal rule that any endeavour must see a return for the 'investor' within a reasonable timeframe, certainly the lifetime of the investor. Who would support, either financially or politically, a mission that would take so long? It seems that we must re-think our conventional ideas of exploration if we are to envisage interstellar travel. And there is a concept that has been a favourite of science fiction writers and space analysts for generations and which offers a system of interstellar travel that contains at least some practicality – this is the concept of the 'ark'.

The idea is that a large spacecraft could provide a home for generations of travellers during a journey that may last hundreds or thousands of years. Although this would require a vast spacecraft capable of sustaining a large number of crew and passengers for perhaps millennia, it would not need to travel particularly fast. If a mission was designed to take 1,000 years, a distance of fifty light years could be covered at just 5 per cent of light speed (100 million kph).

But even if we accept that the technological difficulties of designing and constructing such a craft might be overcome given a suitable level of technology and enough time, there are still considerable drawbacks to the hypothesis.

First, we return to the problem of time-scales. To our minds, the only likely reason a civilization would embark upon such a mission would be to escape catastrophe – an Ark in the Biblical sense, with a proportion of the population finding salvation aboard a spacecraft which then heads off to find a new home. But the early generations of aliens aboard the Ark would have no hope of seeing a new world and would be kept going only by the knowledge that their distant offspring will make it to a distant planet, a proposition that would require a highly developed sense of altruism. It is conceivable that such a scheme might be favoured by aliens possessing a psychological make-up very different from our 21st-century version of human nature. They might think in a similar way to ants or bees and have an in-built instinct for the community rather than the individual. But without an entrenched set of ideals rather different from ours, this project would be difficult to sustain.

However, the most persuasive argument against the vision of an Ark lumbering (relatively) slowly though space is the notion of the 'speed exponential curve'. Using our own technological development as a paradigm, it is clear that the maximum speed at which we are able to travel has increased exponentially over time. For the first 100,000 years of human social development, the highest speed we could reach was about 20 kph – the pace of a sprinting hunter. This was more than doubled some 4,000 years ago when humans domesticated the horse. By the late 19th century another doubling had occurred when trains and motor vehicles were first used. With the advent of propeller-driven aircraft and then jets, the maximum speed potential was multiplied a further three or four times during the subsequent half century. Finally, it increased even more rapidly during the second half of the 20th century with the earliest spacecraft.

At this rate, the speed exponential curve predicts that it should be possible to reach 1 per cent of the speed of light by 2070 and 5 per cent by 2140.

The consequence of all this for our Ark voyagers is that they may well arrive at their destination only to find the planet already colonized by their own race who travelled there in a fraction of the time!

A variation on the idea of the Ark is 'gradual colonization', the notion that an advanced race could 'planet-hop'. This concept is based on the way the South Sea Islanders spread across the Pacific Ocean by island-hopping and consolidation. To see how long it would take for a race to colonize the galaxy using this method we need to consider two things: the time required to make an interstellar journey, and the time needed to establish a colony and to prepare for the next hop.

A conservative estimate for typical sub-lightspeed interstellar journeys would be in the 1,000–10,000 year range, and a reasonable period for colonizing and consolidation would perhaps be one hundred generations. Putting the numbers together we find that the galaxy can be colonized completely in a surprisingly short time, because the growth of the 'empire' would be exponential. If we say the journey time and the colonization time together equal an average of 10,000 years, in an average galaxy in which there are about one billion suitable planets, an advanced intelligence would populate the entire galaxy in less than a million years. Perhaps such a colonization process is what lies ahead for Homo sapiens.

But whatever way you look at it, the possibility of interstellar travel seems remote. Each method is either too slow, too expensive, or both. The best we can hope for is to develop antimatter drives that reduce the time for a relatively short journey to practical levels for the crew but which also forces us to reject the possibility of an organized project with any form of command structure or communication with 'home'.

But perhaps we are being too conservative in our scientific thinking. There may be other ways around the problem using what we might call 'exotic physics'; that is, concepts that, while they do not break the laws of physics as we understand them, certainly stretch the rules.

One piece of exotic physics is the concept of the wormhole. I described this in Chapter 1 as a phenomenon which, under very special circumstances, might be employed to travel back in time. But, as I mentioned, wormholes were originally conceived of as a possible route to interstellar travel, when Kip Thorne considered them as a way in which the aliens in Carl Sagan's novel *Contact* could travel between the stars.

The reason wormholes could be useful to interstellar travellers may be understood if we remind ourselves of how they might be formed. If you recall, in creating his general theory of relativity Einstein introduced the notion of space-time. This we may visualize as a stretched rubber sheet; if we place a heavy ball in the middle, the sheet around the ball will become misshapen. This is how we should visualize the warping of space-time around a massive object such as a star. Black holes, which are incredibly dense collapsed stars, curve space so much that within them lies what is called a 'singularity', a point at which the curvature of space-time becomes infinitely sharp and all the laws of physics break down.

Wormholes are created when two singularities 'find' each other and join up, creating a potential shortcut for interstellar travellers which bypasses the need to travel between point A and point B using the conventional route.

Naturally, this is a very attractive concept, one that might eliminate at a single stroke all the problems faced by near-lightspeed travellers. But, of course, utilizing wormholes is far more easily said than done; and indeed, it may prove to be impossible.

First, wormholes are still pure speculation. They are not disallowed by the known laws of physics, but neither are they certain to exist. And even if they do exist, they would probably be quite rare. The second problem is that, until they were used, it would be impossible to know which parts of the universe they linked. Furthermore, even if they were usable, they would offer only a very limited service, linking two specific points in the universe. It would be a bit like having a motorway connecting London with some other mystery location with no junctions or turn-offs en route.

Ignoring these drawbacks, what do we know about the physical nature of wormholes? Well, we know that black holes are not nice places. We know that a black hole possesses such a powerful gravitational field that any matter that comes too close is sucked in and broken down into a soup of fundamental particles and energy. A wormhole is a close relative of a black hole, so it would be fair to assume that a naturally occurring wormhole would offer any intrepid travellers a very bumpy ride indeed.

Wormholes may provide a means of transport for a sufficiently advanced civilization even if such a thing seems practically unimaginable for us in the 21st century. There is, though, one other alternative which might facilitate a way around the lightspeed restriction. This is the 'warp drive', a concept made famous by the TV series *Star Trek* and the means via which the citizens of the galactic empire of Isaac Asimov's *Foundation* stories traversed interstellar distances as easily as we travel the underground.

Asimov and others have described their interstellar travellers journeying through 'hyperspace' and using 'space warps'; but few authors have attempted to explain the concept in any detail. The technique of hyperspatial travel has been imagined as the only possible way to circumvent the impracticalities of sub-lightspeed travel and the nuisance of having to work within the laws of physics. But what actually is 'space-warping', and how could a spaceship travel through 'hyperspace'?

Another name for warping could be 'surfing', because the idea is based upon the principle of manipulating space-time itself so that the space vehicle rides a 'wave'. The spacecraft would have the ability to alter space-time, so that it expanded behind the craft and contracted in front of it. This means that even though the craft is itself moving relatively slowly, the departure point would be 'pushed' back a vast distance and the destination 'drawn' nearer.

This sounds like a cheat, but again there is nothing in the laws of physics that precludes it. For the system to work, however, space-time would have to be distorted significantly (or else the effect would be so small that sub-lightspeed travel would probably be quicker). The difficult aspect is thus once again the amount of energy required. Observation of our Sun shows that its mass curves space-time so that it bends light by just one thousandth of a degree. For a spacecraft to propel itself via the expansion and contraction of space-time it would have to distort the space-time continuum far more than this. In some respects the vehicle would have to behave a little like a tiny black hole. Using this as a basis for calculating the energy requirements, the result sounds depressingly familiar. To make a black hole the size of a typical spaceship, say a disc 100 metres in diameter, we would need a mass of about 100,000 Earths compacted into that space. Expressed in terms of energy this would be about equal to the entire output of the Sun during its lifetime.

What, then, may we conclude about the chances of any intelligent beings developing the means to travel interstellar distances? All forms of sub-lightspeed travel are restricted and the options for circumventing the lightspeed barrier appear to present insurmountable technical difficulties.

It may well be that extraterrestrials have developed ways of producing enough energy to distort space-time or to manufacture usable wormholes. Alternatively, it is possible they have conceived of methods of interstellar travel that are presently beyond our understanding. To do this, their technology would need to be many thousands of years in advance of our own. But, as I have mentioned earlier, this is certainly not unfeasible given the age of stars and the very different rates of evolution possible on other planets.

Some sceptics hold the very narrow-minded conviction that because the difficulties associated with interstellar travel make it appear an impossibility to us it must therefore be impossible for any other civilization. Clearly, though, if a culture can survive long enough they will develop the skills needed to do almost anything within the laws of physics, including the ability to free themselves from the shackles of their own local environment and to go exploring. In view of this, it would do us no harm to bear in mind what Arthur C. Clarke once said on this matter: 'Any sufficiently advanced technology,' he declared, '. . . is indistinguishable from magic.'

5

Superganglion

Are Telepathy and Telekinesis Possible?

Thurlow is a fine fellow, he fairly puts his mind to yours.
                                                    Samuel Johnson

According to the background story of *Doctor Who*, the Time Lords possess telepathic and clairvoyant powers. These are facilitated by a special part of their brains called a 'superganglion', which permits them to create a 'race consciousness', an invisible net that allows telepathic communication. This can also be used to produce a uniquely powerful melding of their minds to form a *gestalt*. On other occasions, a Time Lord may even mentally communicate with their Tardis.

However, when a Gallifreyan becomes an outcast or renegade, such as the Doctor, the telepathic link is dramatically weakened, so our hero actually has only very limited mental powers, which he may draw upon in emergencies to protect his companions. But how much of this is pure fantasy? How seriously are telepathy, clairvoyance and other mental powers taken by science?

The idea of telepathy is certainly an ancient one, probably as old as civilization itself. But during the past hundred years countless experiments have been conducted in an effort to pin down the phenomenon and to attempt to explain how it might work. These experiments have become gradually more refined, and researchers claim to have eliminated almost every possible way in which the phenomenon could be faked; the scientific community, however, still sees such tests as little more than elaborate tricks.

77

The reason for this scepticism is that telepathy is an elusive phenomenon that never seems to work properly in the presence of a sceptic. This is important to the scientist, because one of the central tenets of science is 'repeatability'. If a scientist claims to have observed a physical phenomenon and conducted experiments to measure the effect, it is only taken seriously by the scientific community if the effect can be repeated under identical conditions by other scientists. If the experiment is unrepeatable, serious doubt is cast upon the original evidence.

But, before we dismiss altogether the concept of telepathy as nothing more than the product of overactive imaginations, let's consider what believers in this ability actually claim.

First, what do we really mean by telepathy? The image from science fiction like *Doctor Who* is of a being with the power to look into another's mind and to pluck out thoughts as they wish, or to manipulate the thoughts of their subject, to make them do things against their will. This, though, is a rather extreme form of telepathy. On a more realistic level, it may well be that all of us are capable of a type of telepathic experience based not upon supernatural abilities but upon enhanced perception.

The psychologist James Alcock has investigated this matter and has described an effect he calls 'backward masking'. An everyday example of this occurs when you are thinking about someone you may not have seen for years and then you turn the next corner and there they are, crossing the street.

The initial response to such an event is surprise and a sense that perhaps you have just experienced something so unusual it must have a supernatural explanation. But what has really happened is that you have noticed your old friend *subliminally* before you turned the corner.

Such subliminal effects are well known to psychologists and have been validated by laboratory experiments. If a subject is shown an image for about a tenth of a second, they will be able to recall some of the features of the image, but if they are shown a tenth of a second flash followed by another image lasting longer, the first is forgotten although it can influence the reported description of the second. For example, if a picture of a man is shown for a tenth of a second followed by a different man holding a knife, the subject's description of the knife-wielding figure is influenced by the characteristics of the first subliminal image.

For the majority of people, though, this is a rare event, hence the feeling that something supernatural is going on; but the ability to subliminally register observations might help to explain a natural process that has recently been dubbed 'ultra-sensory perception', or USP.

We are all familiar with the idea of body language or non-verbal communication, but few people use it consciously. Facial expressions, head movements, body positions, tones of voice and even odour can all send us subliminal signals, and often the interpretation of these is subconscious. Head and facial movements give the most information about the type of emotion being expressed by another person. We are all instinctively able to interpret these, but politicians and actors are trained to pick up more subtle signals.

This ability employs nothing more than our regular five senses, but, instead of the information being processed by our conscious minds on full-alert for signals such as those we receive when we are concentrating on something, the images are filtered into other regions of the brain and siphoned off without our knowing it. Often they are only interpreted later. And our natural senses can sometimes surprise us with their sensitivity.

Recently, the term 'cocktail party syndrome' has entered the vernacular. This is the buzz phrase used to describe everyone's enhanced receptivity to their own name. Above the general hubbub of a cocktail party or some other noisy environment, we can pick out our name, even if it is whispered from the other side of the room.

The cocktail party syndrome is simply a survival mechanism left over from early human development. If our name is mentioned it means we might be called upon in some way. It may signal the approach of an aggressor or a rival attempting to identify us. Alternatively, our name could be mentioned for a reason that is beneficial to us, something we do not want to miss out on.

This ability to subliminally notice things of which we are not consciously aware is the result of a filter system. If we were to give equal importance to everything we picked up with our senses, we could not focus on what is important, and this would be a hindrance.

To overcome this our brains are programmed to know what is important and what is not, and we can then grade these sensations.

It is quite possible that modern humans are not as adept at interpreting subliminal messages as our ancestors once were. The reason for this is that, because of the culture we have established, we have developed certain skills to a very high degree but at the same time neglected others and pushed them into redundancy.

Our ancestors, some suggest, used hand signals and other forms of body language before verbal communication was developed some 50,000–75,000 years ago. One obvious reason verbal communication was developed and adopted over sign language is that it frees the hands and does not need a visual element. If a hunter is trapped by a wild animal as he returns to the tribe, he can call for help even while clinging to the branches of a tree as the animal circles below; sign language does not offer this versatility.

Sadly, although we have retained a little of our ability to read body language and to sense a range of signals detected using our sense of smell, most human cultures have not been able to evolve a twinned path of development, keeping in conscious touch with our primitive instincts but at the same time adopting the sophistication of language. But although we do not notice them, we still possess a pale memory of these skills. Some are better than others at using them, and it might be possible to train people to retrieve these lost talents. Imagine the sensory abilities that specially trained individuals could use if their natural talents and their ability to communicate better with their unconscious facilities were developed to the equivalent of, say, Olympic standard. Such people would be seen as genuine telepaths, even though they would be using nothing more than the senses we all possess.

Although the Doctor is a Time Lord and possesses some natural telepathic ability, he is also able to utilize a talent for subtly detecting and interpreting body language and behaviour. Such talents are actually more common than we might think. Consider for example the extraordinary sensitivity of the wine-taster's palate, or the link between the fingertips and the brain of a blind person reading braille, or of the musician with perfect pitch.

There is also the possibility that if the human body is placed in an unusual environment in which the range of signals is extended beyond what we are used to, we might utilize aspects of our senses we did not realize we had. When astronauts were first sent into space during the early 1960s they reported seeing strange flashing lights. One explanation for this was that they were seeing images created from light just a little outside the range of the electromagnetic spectrum they experienced visually on Earth. The eyes of the astronauts detected the images, but their brains were untrained in interpreting the information.

Some relatively simple animals also display skills that appear on the surface to be paranormal. Dog fish catch flat fish by picking up tiny muscle movements of their prey, hiding invisibly beneath the sand of the seabed, and some species of eel are surrounded by a 'sensory net', a form of electromagnetic field that can detect the presence of other creatures within range, a little like radar. In reality, there is nothing supernatural about these abilities: they are the result of particular evolutionary paths along which these animals have developed.

Within human communities, some of these skills are still to be found in a limited sense. The Cuma Indians in the San Blas Islands off the coast of Panama are said to use odour as a way of helping to judge each other's moods, and they clasp each other under the arm pits and then sniff their palms when they meet. Our sanitized version is to shake hands, but people from developed nations are also subliminally sensitive to smells, and these can influence our feelings towards others without us realizing it.

We all have sensors in our joints and muscles which tell us where we are in three dimensions, and there are other sensors in our inner ears that pass information to our brain about the force of gravity and our own movement. We also have elaborate systems within our bodies that regulate temperature, monitor the level of chemicals and control our highly sophisticated metabolism. Perhaps there is no need to look beyond these to interpret most telepathic experiences.

The difference between all of these sensory effects and telepathy is a question of scale. All forms of ultra-sensory perception, from differentiating between closely associated smells and flavours to the ability to register an image lasting only a tenth of a second, are measurable responses. If genuine telepathy is possible it must operate in one of two ways. Either it uses an extreme form of an information transmission system with which we are already familiar, most probably a region of the electromagnetic spectrum; or else it employs a completely unknown form of information transfer. If the former, the reason we have not been able to detect it is perhaps because our instruments do not operate within the necessary range or they are not sensitive enough. But if the latter is true, we may never develop machines with which we could detect or measure the effect, at least not until this alternative means of conveying information is understood. Marconi, the inventor of the radio, would not have thought of constructing his prototype without being aware that radio waves existed. If he had built a radio in ignorance, finding what to him would have been hypothetical radio waves would have been a rather arbitrary business.

The human brain contains about ten billion neurones or nerve cells, any one of which may have many thousands of connections to other cells, making it the most complex machine known to humanity. Each neurone acts like a binary gate in a computer, switching on or off. In this way, thoughts, emotions, decisions and inspirations are formed and transmitted through a vast network. The neurones are linked by axons, the tips of which do not actually touch. Instead, a signal is passed along the axon and crosses what is called a synaptic gap to another neurone, the process taking about a millisecond (one thousandth of a second). This impulse has an electrical potential of about 120 millivolts and is produced by chemical means, using charged atoms called 'ions' which are triggered to fill the synapse and make the connection to a neighbouring neurone. It has been suggested that if telepathy is possible, the mechanism by which thoughts are transferred has to be explained at least to this level of the process.

One idea is that some form of 'leakage' occurs during the countless individual steps that constitute a thought, and that a telepathic individual can somehow pick up this leakage and translate it into meaningful images. This would be a little like using a phone-tapping device to listen in on someone under surveillance. That is, except for one very important difference: the level of complexity involved in the human brain is several orders of magnitude greater than in the most powerful phone exchange.

Phone-taps work on the elementary principle of siphoning off a signal travelling along a single, relatively large wire or using a remote receiver to access a signal sent between two individuals. By this analogy, the telepath (the mental phone-tapper) would access a single impulse between two neurones. But this would of course be quite useless, because the simplest thought or instruction requires many thousands of neurones working in unison, and a single 'neurone tap' would gather nothing of any value.

Perhaps a telepath could tap into a multitude of neurones simultaneously? The difficulty with this idea is the decoding process. How would the telepath's receiving equipment manage to decipher all the trillions of impulses racing through the brain at any given moment, and disentangle the information they seek? They may want to learn what their subject is thinking about a particular issue, but they would also receive signals passing on instructions to release enzymes, to scratch a leg, to control the bladder, and so on. Continuing with our phone-tapping analogy, a telepath would be tapping into the most sophisticated telephone exchange imaginable, trying to pluck out a few tens of millions of related conversations simultaneously, then piecing them together to make a coherent message. Of course, different regions of the human brain are responsible for different functions, so if the telepath could tune into certain regions then the task might be a little easier, but clearly the prospects are not promising.

An alternative suggestion, put forward by some parapsychologists, is that during the process of impulse transmission the brain releases hypothetical particles called 'psitrons'. Although supporters of this idea speculate that these particles would be released in large numbers, they have not yet been detected. According to enthusiasts of the theory, this is because psitrons possess no mass or energy.

Although this sounds ridiculous, the notion of similarly ethereal particles is not without precedent. In the early days of quantum mechanics, physicist and Nobel laureate Wolfgang Pauli predicted the existence of chargeless, almost massless particles called 'neutrinos', which were eventually observed in 1956.

The psychologist Carl Jung collaborated with Pauli on a book exploring the paranormal entitled *Interpretation of Nature and Psyche*, and the two men maintained an open-minded approach to the possibility that telepathy could be explained by resorting to the esoteric fringes of known physics, asserting that 'the microphysical world of the atom exhibits certain features whose affinities with the psychic have impressed themselves on physicists.'

But this is a terribly vague statement. It is easy to draw a hypothetical link between two disparate subjects like psychic phenomena and physics in this way, but it does not address the key facts. The crucial differences between the neutrino and the hypothetical psitron are that the former fits perfectly into the family of known particles, that it plays a recognizable role, and that it was predicted by the strict mathematics of quantum mechanics before it was detected. Anyone can think up an imaginary particle to explain a phenomenon, give it an appropriate name and suggest that it lies at the root of an unprovable process. Believers have even gone so far as to suggest that psychic powers do not work in the presence of sceptics because the wills of the doubters suppress the action of these particles.

Psitrons have never been detected, but we must accept that this does not mean they do not exist. It is possible that some form of field or resonance or even an array of particles are produced as a by-product of brain activity, but until these are found and their properties understood they should be considered as pure conjecture.

The brain does of course produce measurable potentials which are associated with different brain states, and their discovery by Richard Caton in 1874 raised hopes that science had stumbled upon the method by which telepathy might work. The reality is sadly much more mundane.

Four distinct types of rhythm have been identified in the human brain, and these correspond to different brain states. These waves are due to electrical activity and are manifested as oscillating electrical currents. They may be detected by a machine called an electro-encephalograph (an EEG), which picks up the tiny electrical impulses through electrodes attached to the scalp and amplifies them. The signals can then be represented out on a graph.

The four distinct brain waves are placed in frequency bands and measured in cycles per second or Hertz (Hz). When the brain is resting and relaxed, it produces 'alpha rhythms', which are detected between 8 and 14 Hz. 'Beta rhythms' correspond to activity and predominate when the brain is working, solving problems or facilitating movement such as walking, or running. These rhythms occur between 13 and 30 Hz. At the other end of the brain-wave spectrum are 'delta waves', which are produced during sleep. These have widely spaced peaks and oscillate at between 1 and 4 Hz. Finally, 'theta rhythms', which are produced when the brain is in a deep sleep or a trance state, resonate between 4 and 7 Hz.

Brain waves captured the imagination during the 1970s and the public was inundated with devices said to artificially induce relaxing alpha rhythms. In reality, these merely exploited what yogis and Zen adepts had known for a long time, the fact that individuals are able to control their own brain waves. It led to an awareness of what is now thought to be a fourth state of consciousness, a deep relaxation state or meditative condition corresponding to theta-rhythm production.

EEGs are used extensively in psychiatric treatment and are particularly useful in the treatment of epileptics, who exhibit disrupted brain wave patterns. But though electrical impulses detected in the cerebral cortex (the outermost layer of the brain, a few millimetres thick) reflect the overall brain state, they cannot be deconstructed in order to draw off particular thoughts or even emotions. Research into brain waves has generated benefits for medicine, but it has not unveiled the source of apparent telepathic ability.

Some researchers, though, claim that certain rhythms are more pronounced when subjects are believed to be acting telepathically. These results are based upon the use of EEG machines during telepathy tests that show that alpha rhythms accompany supposed 'thought transference'. But this is misleading, because alpha waves are most noticeable when an individual is in a relaxed state, which is also the brain state most clearly associated with telepathic successes in laboratory tests.

Putting aside attempts to find particles or wave-forms to explain telepathy, some parapsychologists have suggested that the telepathic experience is a holistic effect, some form of response to a network made up of all human consciousness. It may have been this concept that one of the founders of quantum theory, Erwin Schrödinger, had in mind when he said: 'I – I in the widest meaning of the word, that is to say, every conscious mind that has ever said or felt "I" – am the person, if any, who controls the "motion of atoms" according to the Laws of Nature.'

The way in which a 'human network', analogous to the idea of the Gallifreyan mind link, would work, or a network that includes all living beings could operate, remains little understood. Some researchers have made an attempt to clarify the concept, or to link it with aspects of biology and psychology, but the results have caused only further controversy and, in some cases, confusion.

A related phenomenon is a process called 'formative causation', which was first postulated and popularized by the British biologist Rupert Sheldrake in his book *A New Science of Life*, published in 1981. In essence, Sheldrake suggests that systems 'learn', or that it is easier to repeat something if it has already been done. The mechanism for this is called 'morphic resonance'.

Initially, this sounds like a rather vague notion, and Sheldrake has come in for intense criticism from many other scientists, but he has spent the past twenty-five years conducting experiments which he claims verify the concept repeatedly.

One of Sheldrake's demonstrations of the principle is based upon linguistic patterns. He asked a Japanese poet to send him three similar verses. One was a meaningless string of words, the second a freshly composed verse, and the third, a well-known poem familiar to Japanese schoolchildren. He then showed the three pieces of writing to a group of Westerners, none of whom could speak any Japanese. What he discovered was that all of the subjects found it far easier to memorize the traditional poem than the other two.

His conclusion was that the traditional poem had somehow become ingrained into human consciousness via morphic resonance. Based upon this and numerous other tests, some involving living beings as well as experiments involving inanimate matter such as growing crystals, Sheldrake and his supporters believe that all things resonate with their own kind: 'like resonates with like'. In other words, there is a network of human interaction, and similar 'fields' around other species and even other inanimate objects.

Perhaps this is a weak form of what the writers of *Doctor Who* suggest to be the link between Gallifreyans that allows them to form a mental collective or 'superbrain'. But critics of Sheldrake argue from the opposite perspective. If we return to the example of the Japanese poem, could it not be that it had survived and become familiar to schoolchildren because it was easy to learn? Could not certain functions become easier with repetition because those doing them have a natural affinity towards them and avoid tasks that do not come naturally?

If morphic resonance is a real phenomenon, it may point the way to alternative ways in which minds communicate. The traditional image of telepathy is that it occurs via pseudo-physical means, facilitated perhaps by rays or particles, but the truth may be far more subtle. In a sense, any artist may be communicating with their public telepathically by using morphic resonance to cross the barriers of space and time. Perhaps there are special people (and animals) that have a greater sensitivity towards these resonances than we usually experience.

Whatever mechanism lies at the root of the telepathic experience, parapsychologists have been obliged to follow the traditions of science in attempting to demonstrate psychic phenomena. This is really the only way in which they can hope to convince a sceptical scientific community and to develop an understanding of what is happening, if anything.

Researches into paranormal phenomena began during the nineteenth century, but the subject really came of age with the American parapsychologist Joseph Rhine, who summarized his findings in his 1934 book, *Extrasensory Perception*. Rhine worked at Duke University in North Carolina and pioneered the use of what became known as Zener cards (after their inventor, Karl Zener).

There are five designs on the cards: a circle, a square, a star, a plus sign and three wavy lines. Experiments involve the investigator taking the top card from the pack and attempting to transmit the information on the card to the 'reader' or subject, to see if they can identify the symbol on the card. According to probability, there is a 20 per cent chance of simply guessing correctly, but in some trials subjects obtained remarkably high scores. On one occasion a participant got

588 'hits' from just over 1,800 trials – a success rate of 32 per cent. This does not sound like a great improvement on the average, but the likelihood of achieving such a score by chance is astronomically high. In a variation of the tests, Rhine offered a subject $100 for every success. They produced a run of twenty-five successes, netting $2,500, a result calculated to have odds of just under three thousand million million to one against.

After Rhine's experiments hit the headlines, other researchers followed his lead and rapidly brought the study of parapsychology into disrepute with a series of infamous fakes. Since then, parapsychologists have expended great efforts in attempting to develop fraud-proof experiments to demonstrate what they believe to be a genuine and measurable phenomenon.

Modern experiments rely upon random number generators which churn out numbers which are supposed to be completely without pattern, a little like the lottery machines wheeled out each Saturday evening. These experiments are seen as more reliable than the early Zener card tests, but a new generation of researchers are determined to create experiments that are as sophisticated as possible and comply with the necessarily strict guidelines of science.

The latest experiments are known as *Ganzfeld*, or blank-field studies. These involve participants undergoing sensory deprivation in a form of isolation tank. Ping-pong balls are placed over their eyes and white noise is played through headphones. The experience has been compared to staring into a formless fog. After about fifteen minutes most subjects experience hypnagogic images, the sort of image often experienced on the edge of sleep. A sender, usually a friend or close relative of the subject, is placed in an acoustically shielded room from where they try to send an image, usually a one-minute long visual sequence or a static image.

This research is being conducted at a number of centres around the world, including a site at the University of Edinburgh founded in 1985 with a bequest from the Nobel-prize-winning author Arthur Koestler, who was a great believer in paranormal phenomena. So far, there has been little in the way of conclusive evidence to support traditional ideas of telepathy. Like Rhine during the 1930s, the teams have found rare individuals who have achieved impressive scores which lie far outside the normal range of probability. Unfortunately, these results are usually unrepeatable and therefore cannot be deemed in any sense conclusive by orthodox science.

One of the most striking implications from the vast number of experiments conducted during this century is the idea that telepathic ability can apparently be enhanced by a wide variety of factors. The example of the man who produced twenty-five hits in a row when given the incentive of financial reward is a mundane example. Experimenters have become interested in the idea that if other senses are suppressed then telepathic powers can come through more readily. This is the reason for isolating the subject in the Ganzfeld experiments, but it has also formed the basis of experiments linking sleep with telepathy. Once again the results show a small number of impressive, but unrepeatable, events suggesting more pronounced telepathic ability if the other senses are dampened.

Other anomalies could have an influence upon telepathic power. It has been found that children with mental defects score higher in telepathy tests, and in one set of experiments a child known simply as 'the Cambridge boy', who had been born with physical and mental disabilities, achieved well above average scores when his mother was present.

The explanation for this is that if telepathic abilities really do exist then they may be more useful to individuals who cannot communicate in the conventional manner or have their other senses suppressed in some way. There have also been a number of unconfirmed cases of telepathic powers becoming apparent during life-threatening situations. These have been dubbed by parapsychologists as 'need-determined' cases, or 'crisis telepathy', but are dismissed by orthodox science as apocryphal.

According to some researchers, this form of telepathy could be explained by its survival value and might even be a genetically favoured trait. Humans, they argue, have submerged this talent under other more readily developed and utilized abilities, but some rare individuals are more in tune with this power and it comes to the surface in an emergency.

It has also been supposed that other species display this facility. During the 1970s Soviet parapsychologists attempted to demonstrate this effect experimentally. They took a set of newly born rabbits away from their mother and killed them at set, recorded, times. The mother was wired up to an EEG and her brain patterns monitored. According to official reports, the mother rabbit displayed sharp electrical responses at the precise moment each of her offspring were killed. Unfortunately, because news of this experiment merely leaked out from Soviet Russia and was not officially reported, it is difficult to verify, and nobody in the West has so far re-investigated it.

Sceptics continue to pour cold water on the entire phenomenon of telepathy. One of the most usual arguments is to ask why telepathic individuals do not use their skill to win lotteries or to chalk up staggering success at the race track? They also wonder why in lab tests the talents of the claimants mysteriously vanish.

The problem with telepathy is that a century of investigation has turned up little evidence that complies with standard scientific practice. But, during this same period, science has performed quite apparent wonders, from curing diseases to placing humans on the surface of the Moon. But a study by the highly sceptical National Research Council in the United States found in 1988 that there were what they called 'problematic anomalies' in some experiments that could not be explained; in other words, incidents of success that could not be accounted for merely by chance. And, despite the apparent lack of evidence, many people believe in telepathy. In one survey, 67 per cent of people questioned said they had experienced ESP.

Psychologists have noticed that 'a sense of deep personal conviction may be the key to achieving good results,' but this is surely not the whole story. There is still no satisfactory scientific explanation for what supporters claim has occurred during a growing number of experiments, but it would be unscientific to conclude from this that telepathy is imaginary. It may just be that we don't know how it works.

Telepathy though is really only the tip of the psi powers iceberg. If we think of telepathy as being 'mind communicating with mind', then what are the chances of mind interacting directly with the physical world, with matter?

Psychokinesis, or PK, is defined as 'The apparent ability of humans to influence other people, events or objects by the application of will, without the involvement of any known physical forces'. Whereas telepathy could be described as an interaction between two psychic 'fields' or 'forces', PK involves an interaction between the mental and the material, so it is one stage further down the psychic route. To the sceptical, this looks like a marriage from hell, but from the perspective of parapsychologists it is merely a natural progression from the more prosaic telepathy.

PK has been the subject of even more experimental work than thought transference, and according to how you view this evidence, it has either been proven beyond doubt to be a genuine, natural process or else all the tests and experiments conducted during a period of almost a century have been faked or may be explained by other factors.

The earliest serious attempt at trying to quantify the concept of PK is attributed to the parapsychologist J. B. Rhine, who worked on telekinesis experiments concurrently with his attempts to pin down telepathy during the early 1930s. He was led into the subject when a young gambler told him he could influence the fall of a dice by will-power alone. Rhine immediately set about conducting thousands of tests on scores of subjects, in an attempt to reach a statistically meaningful conclusion that would show if there was any such effect.

He based the test upon participants trying to will two die to produce a score of more than seven. As with the results of his telepathy tests, Rhine and others found that most people achieved scores that deviated little from the values expected by chance, but once in a while he turned up individuals whose scores did not fit the normal pattern. On a few occasions, he found someone whose scores deviated from the average to such a degree that the probability of it happening by accident was sometimes placed at millions to one against.

In one set of experiments Rhine conducted 6,744 tests. These should have produced 2,810 successes by chance alone, but one subject achieved a score of 3,110, a deviation calculated to occur by chance only once in a billion tests.

Rhine's experimental methods attracted criticism almost immediately, but, as with his telepathy experiments, he went to great lengths to de-bug his tests of any chance of trickery or unintentional influence.

After nearly thirty years of such experiments, he reached the conclusion that 'The mind does have a force that can affect physical matter directly.'

Since Rhine's seminal investigations, literally hundreds of groups around the world have conducted other forms of PK tests. During the 1970s researchers led by Helmut Schmidt at the Mind Science Foundation in San Antonio, Texas, replaced Rhine's die with a Geiger counter. The reason behind this was that Geiger-counter readings derive solely from radioactivity produced by the breakdown, or 'decay', of radioactive nuclei. This decay process is completely random and as close to fraud-proof as parapsychologists could hope to get.

Schmidt had participants attempting to influence the read-out from the Geiger counter – usually a display showing a series of flashing lights or an oscilloscope screen showing a wave pattern. Variants on this theme were developed in the 1980s using white-noise patterns generated electronically.

One of the leading parapsychologists currently working with PK is Robert Jahn, an engineering professor based at Princeton University in New Jersey. He developed the white-noise experiments and has gone on to try different versions of what he calls a 'random event generator' – a machine that produces random displays or number sequences; the electronic equivalent of tossing a coin thousands of times. The generator is fitted with a collection of safety devices to detect any change in temperature, the influence of external magnetic fields or any physical disturbance such as tilting or weighting the machine in any way.

Ignoring the cynicism and sometimes open hostility of his orthodox colleagues, Jahn has spent the past twenty-five years conducting hundreds of thousands of tests, using over 100 subjects to see if there is any deviation from chance. What he has found is still inconclusive. Taking the collection of experiments and subjecting them to statistical analysis, he has found that there is some effect which he judges to be about a 0.1 per cent deviation from pure chance. In other words, on average, one thousand tests throws up one significant deviation from what was expected.

If this all sounds unconvincing, there are other more worrying aspects to Jahn's experiments. He conducted a large number of trials where the subject was placed a long way from the random event generator in his laboratory in America. Some of the subjects were asked to attempt the test from as far away as Africa and England, but, confusingly, Jahn found that their success rate did not vary at all with distance.

In another set of experiments, he asked his subjects to make their attempts at psychokinetic influence several days before the test and again found that there was no difference in the quality of the results. Others have discovered the same odd anomaly. In a different collection of tests conducted by Helmut Schmidt, he disconnected the random event generator and substituted the 'live' read-out with a recording of the signals from the previous day, but didn't tell his subjects what he had done. He found that, if anything, the results were better than they had been in the usual experiments.

PK enthusiasts have produced a very odd explanation for this. The subjects, they claim, are practising what has quickly been dubbed *retroactive psychokinesis*. This involves them sending thoughts back in time and space to the previous day and affecting the read-out.

Astonishingly, many parapsychologists subscribe to this explanation. Cynics merely call it a nonsensical explanation for an obvious and telling flaw in the entire theory and practice of parapsychology. Yet, there is actually another far more mundane explanation which appears to have escaped both camps. If the results on Day 2 are affected by the subject even though the display from Day 1 was being shown to them, perhaps their minds were interacting with the playback device. If we are trying to demonstrate PK, there is no reason why the subject might not control a machine feeding a fake display any more than they might alter the pattern from a random event generator. After all, the playback device is certain to be some form of tape recorder or digital device and as susceptible to PK as any other material system.

So, putting to one side the views of the sceptics and the whole-hearted believers, what scientific conclusions can be drawn from the vast range of PK experiments?

Enthusiasts point to the rare outstanding anomalies and conclude that something odd is happening and that this must point to proof of

PK. But they seem unperturbed by the fact that those cases which deviate greatly from chance are quite exceptional and very rare. More common is a tiny perturbation from the expected, which may perhaps point to a weak force at work or could be due to a number of other anomalies.

One such anomaly could be stray magnetic fields or electric currents. It has been found recently that some people who live near power cables sometimes experience physical illness and depression. This is thought to be due to the close proximity of the brain to powerful electrical currents. All electrical impulses have associated magnetic fields, and those around carriers such as national-grid power cables interact in some way with the similar but far weaker magnetic fields created by the electrical signals produced in the brain. Via some unknown mechanism, these disturbances manifest as physical and emotional instability. In a similar way, it is possible that magnetic fields some distance from the test centre could have a weak effect on the machinery used for the tests mentioned above. Another source might be electrical interference by leakage from equipment elsewhere in the lab or even beyond the building. Experimenters have tried to negate this anomaly by encasing the test equipment and the subject in a special container called a Faraday cage, which shields them from electromagnetic disturbances.

Such a tiny effect as that observed by parapsychologists could be produced by other natural sources. We all live on a giant magnet. Like most planets, the Earth has its own magnetic field which fluctuates naturally due to movements hundreds of miles beneath the crust. It is also disturbed by fluctuations outside the atmosphere. Sunspots, which are cooler regions on the surface of the Sun, are able to disrupt the Sun's powerful magnetic field and this in turn can alter the Earth's associated field. Such magnetic disturbances might conceivably affect electronic machinery and also the fields each of us produces by electrical activity in our brains and from each nerve impulse passing constantly through our bodies.

Other factors to consider are currents of warm or cool air and minute geological disturbances such as micro-earthquakes. Although these factors are almost undetectable, they could be powerful enough to disturb PK experiments.

These objections may sound pedantic, but if parapsychology is to be taken seriously by scientists, it has to play by the same rules as orthodox science. Many critics of PK and other paranormal phenomena cite lack of care or unprofessional attitudes to research as the most likely source for the apparently impressive results.

The only reasonable conclusion to be drawn from the millions of PK experiments conducted since the 1930s is that if the results do demonstrate a genuine effect produced by the human brain, then the effect is very small. And because it is so small, it is incredibly difficult to measure. Parapsychologists have dubbed this effect *micro-PK*, and there is a growing body of evidence to substantiate an anomaly of this type.

In the mid 1970s the psychologist Gene Glass came up with a revolutionary approach to the study of results from parapsychology experiments. He realised the problem with PK was that the effect was so small (usually in the region of 0.1 per cent over chance) that a large number of results would be needed to show up the anomaly created by any genuine paranormal activity. Furthermore, the smaller the effect, the more results would be needed. He then went on to devise a method for studying experimental results using a technique he called *meta-analysis*.

An analogy would be the effort put into tuning a radio. If a signal is strong, such as, say, a BBC transmission picked up in Southern England, it would be easy to tune into it. If, on the other hand, you were trying to pick up a weak signal from a pirate or independent broadcaster, you might have to spend some time fine-tuning the radio to detect the signal. This fine-tuning is equivalent to conducting a large number of samples or tests.

So Glass's idea was to somehow pool the data from all the tests that had been conducted over a long period. The problem was that the tests carried out since the 1930s were quite different from one another. Some experimenters had investigated the possible effects of PK on falling die or wood blocks, while others followed the altering of light displays or the random decay of unstable isotopes using a Geiger counter. But Glass eventually found that, using suitable mathematics, results from disparate tests can be merged, which means that the parapsychologist has access to a far larger sample – tens of millions of results taken by scores of experimenters over a period of some sixty years.

One of the best examples of applying meta-analysis to PK experiments comes from the work of the psychologist Dean Radin, at Princeton University's Psychology Department, and Roger Nelson, a member of the Princeton Engineering Anomalies Research programme (PEAR). They did not conduct their own experiments but instead tracked down over 150 reports summarizing almost 600 separate studies and a further 235 control studies by 68 different investigators, each of whom had been researching the influence of consciousness on microelectronic systems – experiments where the subject was asked to disturb the workings of an electronic random event generator.

To their amazement, they found the probability of the net result deviating from the normal pattern merely by chance was 1 in 10^{35} (1 with 35 zeros after it).

Again, this result does not say for certain that PK is a genuine phenomenon, but it does suggest there is some factor or combination of factors that alters the behaviour of matter other than by the visible, conventional means. Whether this is the effect of human consciousness or sunspot activity, micro-earthquakes or thermal currents, is another matter.

Those convinced that the aberrations shown up by meta-analysis are of human origin believe the effect is generated by micro-PK and that this phenomenon is with us all the time. They suggest that many incidents we might think of as coincidence are a result of this force. When we drop a book and it opens to exactly the page we wanted; when we look through a filing cabinet and put our fingers straight on the piece of paper we've been looking for; when we pot the black without even looking. These, according to the supporters of micro-PK, are all examples of a subconscious ability to influence the way matter behaves. Furthermore, it is thought that this phenomenon works best when the subject is not deliberately trying to make it work, when the individual is concentrating on something quite different.

We have all experienced moments when things have either gone extremely well for us or extremely badly – good days and bad days. We've all experienced beginner's luck, moments when you can do no wrong. It is on these occasions, the believers say, that micro-PK is working at its best, and these are exactly the occasions when we are the least likely to be trying to make it work.

A psychologist called Rex Stanford, working at St Johns University in New York State, has conducted an interesting variation on the usual PK experiment to illustrate the effects of micro-PK. He places his subjects in a locked room and gets them to perform a series of very dull tasks. In the next room is a random number generator. What he does not tell the subjects is that they can only be released from their task and allowed to leave the room when the generator produces a sequence of numbers that appears only once every two or three days under normal circumstances. Yet, on several occasions, some subjects have managed to get out of the room within forty-five minutes.

The problem with accepting micro-PK is that the force producing the effect is the same as that involved in macro-PK. Any system which allows us to subconsciously control the way a book lands or the movement of a ball in a game of billiards is the same phenomenon that could allow us to move objects at will. They are on the same scale and would presumably operate by the same wave-form, particle stream or other inexplicable force. And, it is not whether these effects occur occasionally or frequently that bothers the scientist, it is how they could occur even once. Because, at the root of the dilemma remains the fact that no form of psychic force has been detected, yet we are asked to accept that the mind can interact with matter – the marriage from hell I mentioned at the start of this section.

To illustrate the problem, let us consider the physics of PK. What are the energies involved and is there any compatibility between what is needed and what could be reasonably generated by the brain?

Let us imagine an experiment in which a subject who claims the ability to perform psychokinesis is asked to move an object along a table using just the power of their mind. Suppose our object weighs 50 grams, the mass of a tea spoon. Now assume the participant is to accelerate the object to the modest velocity of 5 cm per second, and to maintain this velocity for a few seconds. If we add a small contribution from friction, the energy needed to do this comes to approximately 4×10^{-5} Joules (one forty thousandth of a Joule).

This is a relatively small amount of energy, roughly equivalent to that stored in one millionth of a gram of sugar. But equally, to produce even this much energy from a force which has so far remained undetected by any conventional means requires a suspension of disbelief. Consider the figures.

As I mentioned, the brain has associated electric and magnetic fields. Now, we could suppose that the electrical impulses from the action of neurones are responsible for creating the force with which the object is moved. But this field must interact very weakly with the material world, because we cannot pick up the force or any form of tangible interaction with any known instrument. Let us be liberal and say that one thousandth of the power of the electrical impulse penetrates the skull and reaches across space to the object and accelerates it to a speed of 5 cm per second for a short period of say three seconds. The voltage produced in the neurones is approximately 100 millivolts, so this would mean that the human brain would have to produce a current in excess of 0.25 amps to provide the necessary energy. To put this into perspective, a current of little more than half of this (0.15A) passing through the heart would kill a person.

One way around this problem is the suggestion that a group of individuals such as the Time Lords of Gallifrey could work together and create a melding of their minds in order to produce the effect of telekinesis. This would certainly alleviate the difficulties of making the numbers work in the above scenario, but we are back to the improbable ways in which individual minds could communicate at a distance without any form of apparatus.

One of the most telling aspects of PK research is the fact that, like telepathy, psychokinetic effects have no respect for distance. The set of experiments conducted by Robert Jahn at Princeton University, in which subjects were apparently able to alter readouts up to several thousand kilometres from the laboratory with as much ease as when they sat a few metres from the detector, illustrates this point.

Almost all known physical forces operate according to the inverse square law. What this means is that as distance is increased, the effects of a force diminish by the square of that distance. For example, Isaac Newton described in his masterwork *Principia Mathematica*, published in 1687, that the force of gravity operated via the inverse square law.

By Newton's reasoning, if planet A circles the Sun at a given distance, it will experience a certain gravitational attraction towards the Sun. If planet B with identical mass orbited at double this distance, it would experience only a quarter of the gravitational attraction experienced by planet A (the inverse of 2 squared). An identical planet C, orbiting three times as far away as planet A would feel only one ninth the gravitational attraction experienced by A (the inverse of three squared).

All known forces operate by this inverse square law, but supporters of PK argue that there are other forms of energy transmission which do not. The most popular argument is that the intensity of radio signals does not weaken very much with distance. From this enthusiasts draw an analogy between radio and 'thought waves'.

The first part of this statement is true. If radio signals are generated by a powerful enough transmitter and fine-tuned using what is called a signal-optimizing system, their intensity does not diminish to any large extent over reasonable distances. But, this is not really the point. Radio signals convey information, they do not move objects or enable the bending of spoons; some form of force is needed to do these things, and all known forms of force operate by the inverse square law. Furthermore, if PK operates by electromagnetic radiation, of which radio signals are a small part, where in the spectrum is this radiation to be found? It might operate at extreme ends of the electromagnetic spectrum, but, as discussed in Chapter 3, the range has been thoroughly searched and no trace of a psi-wave discovered.

If PK is produced by a genuine form of mental energy or acts via some as-yet-unknown force, then the only conclusion to be drawn is that this force has nothing in common with those that human beings have so far experienced in the universe. Depending upon your viewpoint, this fact either strengthens or weakens the case for a paranormal explanation for PK. To the sceptical, it merely reaffirms the claim that, because PK cannot be explained by recourse to electromagnetism, it must be put down to trickery or sleight of hand. To the believer, it confirms that the phenomenon is not governed by the normal laws of physics, cannot be measured by humans wielding electronic gadgets, and lies above and beyond us. What you decide to believe comes down to a matter of faith in science or faith in magic.

6

Supercivilizations
Do Superraces Exist?

As flies to wanton boys are we to the gods.
King Lear, William Shakespeare

The Time Lords are superbeings, and according to the back story of *Doctor Who* the civilization that dominates the planet Gallifrey has been around for at least hundreds of thousands of years. The Gallifreyans are capable of time travel, of course, but they have also mastered interstellar travel, they can regenerate their bodies, thereby lengthening their lives almost indefinitely, and they have learned to tap the resources of many worlds.

It is easy for us to believe that we are cleverer than we actually are. There is no denying that the human race is an ingenious one, that we have produced great things in a short time; but it would be naïve to believe that we have reached the pinnacle of our capabilities.

Civilization as we recognize it dates back to the Babylonians and the Ancient Egyptians, who flourished some 6,000 years ago. On a cosmic scale this is little more than the blink of an eye. Furthermore, we now realize that technology is advancing at an exponential rate. This means that what we are capable of today will be outstripped very quickly, and we will look back at our present stage of development as perhaps rather unremarkable.

The onward march of any culture is inevitable; a civilization that stops developing can only stagnate, and stagnation leads inexorably to extinction. No civilization can remain in stasis. Of course, the quality of a civilization is not demonstrated solely by the level of its technology. We might also judge a culture by its art and its political systems, by the rights of its citizens and by the way it behaves towards other societies. But for the exobiologist the primary interest is to quantify the level of technology and scientific advance of an alien culture, and this has led them to categorize civilizations accordingly.

According to this system, there are four categories into which all civilizations fall. Type 0 signifies primitive cultures, in which the population is dispersed and there is little or no social structure. Type I cultures (which includes us) are those that have developed to the point where they can exploit the natural resources of a single, home world. A type II civilization would be capable of building machines that can process the entire energy output of their sun. This level of development would almost certainly be associated with the ability to travel interstellar distances. Such cultures may also have developed means via which they could circumnavigate the hurdles presented by the light-speed restriction. A culture that had reached this stage of development would be thousands or perhaps tens of thousands of years in advance of us.

A type III civilization would be millions of years ahead of us and would have developed the technology to utilize the entire resources of their galaxy, an ability which to us would appear God-like but which is actually possible within the laws of physics. It would be wrong to consider such beings as supernatural; their advanced state of technological development would be a consequence of the fact that their species started to evolve a little before us. Although we view the technology employed by a type III culture as almost omnipotent, like us, such beings would have originated in a slurry of single-celled organisms on some far-distant planet long ago. It's just that they have had more time in which to develop.

Of course, at the moment we have no idea whether supercivilizations even exist somewhere out there in the universe, let alone what their culture might be like. As we saw in Chapter 2, the logic of numbers would suggest there are many highly advanced civilizations not too far away in astronomical terms.

But, sadly, we have no clear evidence to support the idea that advanced technologies abound in the universe – we are left merely to speculate. We cannot find traces of ancient advanced races in the stars, and we have no artefacts or recorded messages to help us. But this restriction has not stopped some people giving serious consideration to the notion that our civilization was perhaps seeded by an advanced race of humans, or that alien visitors passed our way many millennia ago and left behind traces of their culture.

One of the most enduring ideas of this type is the legend of Atlantis, and the belief that it is rooted in real history and that the Atlanteans were in some way connected with an advanced alien race. Let us investigate the claims, and see whether here we can find evidence of a superrace.

During the past two centuries literally thousands of books and articles have been written on the subject of Atlantis, and it is easy to understand the fascination. Psychologists would claim that the image of Atlantis holds a mirror to our world. It has a similar emotional energy to the biblical Garden of Eden. For others, Atlantis represents an idealized version of our own future. Another powerful psychological impulse to believe there may be some great significance to this legend is that it makes our history more glamorous, it adds gravitas to human development, linking us with something higher.

In some respects, Atlantis is all things to all people. For some it is nothing more than a myth, to others it is a lost continent that may yet be found. For a minority of people it was a place visited by aliens, perhaps even established by extraterrestrials, who then passed on their knowledge to the Egyptians.

According to some pseudo-historical theories, Atlanteans were the forebears of our technological society, the progenitors of ancient knowledge, keepers of what the alchemists and Hermeticists called the *prisca sapientia*. But what lies at the root of these ideas? Did Atlantis ever really exist? And if it did, what sort of place was it? And who were the Atlanteans of legend?

The story of Atlantis has come down to us from the Greek philosopher Plato, who lived during the 4th century BC. Renowned for his masterpiece *The Republic*, devoted to political structure and the nature of government, he also wrote a pair of dialogues – books in which two or more characters discuss points of philosophy, history and science. These were entitled *Timaeus* and *Critias*. In these texts Plato's teacher, Socrates, holds an imaginary conversation with three of his friends. One of these characters is Plato's maternal great-grandfather, Critias, who had heard a story about a place called Atlantis from his grandfather, Critias the Elder. He had been told the story by his father who had learned of the legend from the great Athenian thinker and law-giver, Solon, who had died 130 years before Plato's birth. Solon claimed to have come by the tale when he visited Egypt, where an ancient priest and guardian of the Temple of Sais had been privy to ancient records documenting all remaining knowledge of the lost continent.

According to these ancient texts, Atlantis was a vast landmass populated by peace-loving demi-gods presiding over a global culture that existed some 9,000 years before Solon's time – about 11,500 years ago. Critias tells the story:

> There was an island opposite the strait you call . . . the Pillars of Hercules, an island larger than Libya and Asia combined . . . On this island of Atlantis had arisen a powerful and remarkable dynasty of kings, who . . . controlled, within the strait, Libya up to Egypt and Europe as far as Tyrrhenia [Italy]. This dynasty . . . attempted to enslave at a single stroke . . . all the territory within the strait.

Critias goes on to relate how the Atlanteans overstretched themselves, and lost sight of their peaceful origins, prompting the gods to destroy them.

> At a later time, . . . there were earthquakes and floods of extraordinary violence, and in a single dreadful day and night . . . the island of Atlantis . . . was swallowed up by the sea and vanished; this is why the sea in that area is to this day impassable to navigation.*

Plato's *Timaeus* and *Critias* also detail the political structure and society of Atlantis. The Atlanteans were ruled by ten kings, descendants of five pairs of twins, the offspring of the union of the mortal woman Cleito and the god Poseidon. The kings met at intervals of five years to make far-reaching and long-term decisions based upon

---

*Interestingly, according to the plotline of the *Doctor Who* story *Time Monster* (Season Nine), the Doctor's greatest enemy, The Master, was responsible for the sinking of Atlantis after he released the 'time-devouring' beast Kronos, which had been captured by the Atlanteans.

a voting system. Their meetings took place in the great capital city, which was surrounded by a golden wall. The city had hot springs, temples, exercise areas and a racecourse.

Originally a noble people descended from the gods, the Atlanteans eventually became greedy and sought to dominate realms beyond their boundaries, invading neighbouring states and oppressing less developed cultures. Finally Zeus became angry and destroyed them at a single stroke: the grand palaces and the golden wall were swept aside and sank beneath the waves as the land of Poseidon's children was deluged by the ocean.

For many historians, the account outlined in Plato's dialogues is an example of a technique he used often, most famously in *The Republic*, expounding his philosophical ideas through morality tales and stories that highlighted ethical issues. With such details as the racecourse, the description of the Atlantean capital sounds suspiciously Greek, which would suggest that the original tale as handed down to Plato has been greatly elaborated and embellished by the author to make it fit the requirements of his own culture and to convey his own ideology. The structure of the story is also an ancient one – the notion of a people acquiring too much power and becoming corrupt before being taught a lesson or, in this case, destroyed by the all-seeing, all-knowing gods.

Plato's pupil, Aristotle, took the view that in writing *Timaeus* and *Critias* his master had expanded upon a kernel of truth in order to create a myth to convey his philosophical teachings and he called Plato's work 'political fable'. But despite the fact that Aristotle became the colossus of teaching throughout the world for at least 1,500 years after his death – his ideas became the cornerstone of philosophy and science until the Enlightenment – there remained a large contingent of people who did not see *Timaeus* and *Critias* as mere fables.

Even by the time of the philosopher Gaius Plinius Secundus, known as Pliny the Elder, who wrote the encyclopaedic *Natural History* four centuries after Aristotle's time, the place of Atlantis in the world of philosophy and history was ambiguous. Some scholars held the view that Atlantis had been a real place, lying opposite the Pillars of Hercules, whilst others saw it as simply a myth.

Plato's writing is the earliest surviving source of the Atlantis story, and all other accounts are based entirely upon it. And, because there is so little to go by, it is almost impossible to judge the political structure or the form of society adopted by the Atlanteans. Consequently, those fascinated with the subject have concentrated on finding the location of the lost continent in the hope that one day an expedition will unearth the great walled city and reveal the secrets of the demi-gods.

Throughout ancient times and indeed until the late fifteenth century the Atlantic was a largely uncharted ocean. Roman and Dark Age historians and geographers described a vast array of mysterious islands and isolated lands throughout the Atlantic, almost all of which turned out to be fictitious. These included the islands of the Seven Cities, the Fortunate Isles, St Brendan's Isle and a mysterious island named Hy Breasil, which remained on mariners' charts until the late nineteenth century.

Plato had described Atlantis as lying beyond the Pillars of Hercules, by which he meant the Strait of Gibraltar, the gateway from the Mediterranean to the Atlantic. It was only natural, then, that this information should lead cartographers and explorers to place the mysterious lost continent of Atlantis somewhere in the Atlantic Ocean. But where exactly?

During the first few decades after America was discovered by Columbus, European philosophers thought that this new landmass might be the remains of the lost continent. In the 1550s, the Spanish historian Francesco López de Gómara believed that some of the features of what was then known of America and the West Indies fitted the description Plato had offered in *Timaeus* and *Critias*, and in his masterpiece *Nova Atlantis* published in 1618, the English statesman and philosopher Francis Bacon placed Atlantis in the New World.

But gradually, as America was explored and mapped, it became clear that it had nothing to do with Atlantis. Yet, the idea that Atlantis was to be found somewhere in the Atlantic Ocean persisted.

One of the most determined adherents of this view was the American author and historian Ignatius Donnelly, who was convinced that the Azores are the only remains of Atlantis above sea level. In his book *Atlantis: The Antediluvian World*, published in 1882, Donnelly propounded the theory that Atlantis sank beneath the Atlantic waves and that a feature called the Mid-Atlantic Ridge, discovered in the 1870s (of which the Azores are the volcanic peaks) was a major geographical component of the continent.

Donnelly went to his grave believing he had solved the mystery of the location of Atlantis, but in the 1960s the study of plate tectonics showed that his theory could not be correct. Plate tectonics describes how the present configuration of the Earth's crust has been produced by shifting plates (large segments of the Earth's surface), which have created features such as mountain ranges and ocean ridges including the Mid-Atlantic Ridge. Rather than this range being the remains of a sunken continent, as Donnelly believed, it was created by the movement of the Earth's tectonic plates in relatively recent times.

Today, Donnelly's ideas are dismissed by scholars and scientists, but many still believe Atlantis was a large landmass in what is now the Atlantic Ocean. Some go on to propose that the lost continent was located close to the West Indies.

Support for these claims emerged in 1968, when a diver living in the Bahamas and known to locals as Bonefish Sam met an American zoologist and keen amateur archaeologist called Dr J. Manson Valentine, who was visiting the island. Bonefish Sam showed him an underwater oddity which he thought might be of archaeological interest.

The anomaly is about a kilometre off Paradise Point in the Bahamas, and it consists of what could have once been a wall – though interpreted by some as a road – made from large stones lying under about six metres of water. The stones, now known as the Bimini Road, are each estimated to weigh between one and ten tonnes, and several dozen of them are aligned to run for about half a kilometre in a straight line before ending in a sharp bend.

A few years after this discovery more ammunition was provided for those who believed Atlantis was once located in this region of the Atlantic. In 1975 Dr David Zink, author of *The Stones of Atlantis*, discovered in the same region what appears to be a block of stone that looks very much like concrete and is certainly man-crafted because it contains a tongue-and-groove joint.

Sceptics suggest that this and other artefacts, including some anomalous marble pillars (marble is not found in the area naturally), are simply relics of shipwrecks. These arguments were strengthened when a building found alongside the Bimini Road and originally thought to have been an Atlantean temple was shown to be nothing more exotic than a sponge store that had been built in the 1930s. This has led sceptics to encourage the theory that the Bimini Road was actually produced by a natural phenomenon, an accepted geological process called 'Pleistocene beach-rock erosion and cracking'. Others tread a middle course, suggesting that the underwater feature is natural but could nevertheless have been used by ancient peoples.

Today, the mystery of the Bimini Road remains unsolved, and it is the battleground for supporters and opponents of the many theories surrounding the reality and location of Atlantis. In 1997 a group of British researchers from the Building Research Establishment (BRE) analysed samples from the man-made block close to the Bimini Road and have concluded that it is made from a form of concrete manufactured by an old-fashioned technique and certainly older than the modern-day Portland cement process, devised in 1820. Just how long before is uncertain, so the block could have been produced in Europe anytime between the 16th and early 19th centuries, or it could be far more ancient.

Using an electron microscope, one of the BRE team, Dr David Rayment, head of the organization's Electron Microanalytical Unit, has found a strip of gold in the concrete block which shows clear signs of having been worked by a skilled craftsman. But, although this is a fascinating development, it gets us little closer to determining the origin of these finds.

Clearly, more research on these materials is needed before a definite conclusion can be reached. Carbon dating would be of little use in this case, because the concrete block is man-made, but one possibility would be to try to find pollen grains or other organic material inside the core of the block which may have been deposited there when it was being produced. These could be matched with samples found in different parts of the world in an effort to determine where the block was made. And, of course, the pollen grains or other natural materials could then be carbon dated.

Among the groups interested in this esoteric subject, the Atlantic Ocean remains the most popular location for Atlantis. But there are alternative theories that have been proposed recently after some researchers considered afresh Plato's original descriptions.

Close inspection of Plato's account shows many confusing contradictions and anomalies. Firstly, it seems that he has exaggerated all the dimensions by a factor of ten. This has been noted by scientists, who point out that in his description of the great capital of Atlantis, the Royal City, Plato ascribes a length of 10,000 stades or 1,135 miles to the city wall. Even Plato questioned the validity of this figure in his transcription, and it does indeed seem excessively large even for a culture ruled by demi-gods. The Great Wall of China is 1,500 miles long, but a wall over 1,100 miles in length would circle Greater London twenty times. If we reduce the length of the wall by a factor of ten we have a more reasonable number.

Plato also says that Atlantis existed 9,000 years before his time, which places it in an era when the rest of the world was still in the Palaeolithic period or Old Stone Age, at least 6,000 years before the origin of Egyptian civilization. It has been argued that the mythical flavour of the lost continent is eradicated if we again divide Plato's figure by ten, placing the high point of Atlantean culture 900 years before Solon, but it fits with the description offered by Plato of battles between the Atlanteans and the embryonic state of Athens during the time the kings of Atlantis were attempting to expand their empire.

This mistake (exaggerating by a factor of ten) could have occurred, it is argued, because the Egyptian copyist mistook the ancient Egyptian symbol for 100 (a coiled snake) for the lotus flower, the symbol for 1,000. This would be analogous to mistaking the British billion (a million million) for the American billion (one thousand million).

A further confusion arises over a simple phrase in the original tale. Plato was told that the lost continent was: '. . . larger than Libya and Asia combined', but the Greek words for 'greater than' and 'between' are almost identical: Plato could have meant Atlantis was located 'between Libya and Asia'.

If for the moment we take this idea to be fact, it places an entirely different complexion on the tale of Atlantis, at once making it more prosaic and allowing the possibility of links with another culture that is known to have existed at the same time, but one far from the Atlantic Ocean – the Minoan civilization of Crete.

Beginning in 1900 with Sir Arthur Evans, a succession of archaeologists have studied the region encompassing the islands of the Aegean lying south of mainland Greece, and these researchers have gradually pieced together a picture of what may have happened to a great civilization that once lived there around 1500 BC.

The most southerly of the Greek islands is Santorini. Today it is actually a collection of three islands, the largest of which is the beautiful isle of Thera, a major tourist attraction with its black sand and crystal clear waters. But 3,500 years ago Santorini was a single, almost circular island. It was blown apart by a massive volcanic eruption thought to be four times more powerful than the eruption of Krakatoa in 1883.

The eruption of Krakatoa has been estimated as equivalent to one million Hiroshimas, and although this may be an exaggeration, the explosion that occurred on Santorini around 1520 BC must have been truly devastating. It is believed to have created 100-foot waves that swept in all directions from the island, entirely engulfing another advanced community living a mere 96 kilometres north – the Minoan civilization that was then thriving on the island of Crete.

Evans, who discovered the Palace of Knossos and unearthed the lost history of Minoan culture, did not link the destruction of this civilization with the volcanic eruption on Santorini; but others did soon find a link. As early as 1909 some scholars were suggesting that the ruins of the Minoans were in fact the lost Royal City of the Atlanteans. By the 1930s, the Greek archaeologist Spyridon Marinatos made the link between the two events after he found further Minoan remains in the north of Crete along with pumice, a frothy form of volcanic glass left over from a volcanic eruption. Later, during the 1960s, ruins of an advanced culture were found on Thera, including the remnants of a massive circular channel on the edge of what remains of the original island. This matches Plato's description of channels circumventing the great metropolis of Atlantis.

Other possible links emerge from a comparison of the Minoan culture and the legends of Atlantis. According to Plato, the Atlanteans worshipped the bull, and it was discovered from the ruins at Knossos that the bull-cult also lay at the centre of the Minoan religion.

If we take Plato's geography as misguided, along with his time-frame being wrong by a factor of ten, then the evidence for superimposing Atlantis on Crete is compelling. This then suggests that the tale of Atlantis that passed from Solon to Plato was an Egyptian legend based upon an event that may have taken place some 900–1,000 years earlier.

However, even this rather neat explanation has its critics. Recent archaeological findings using accurate dating techniques suggest that the volcanic ash from Santorini is at least 150 years older than the date assigned to the destruction of the Cretan palaces, implying that the volcanic eruption on Santorini did not destroy the Minoan culture after all.

Evidently, a great deal more work has to be done on the possible links between Santorini and Crete before a plausible hypothesis linking Crete with Atlantis can be formulated; but for many, what appeared to be a promising connection is too flawed to be accurate and they are actively searching for new solutions.

One of the best-researched alternative theories of recent years has been the work of the writer Graham Hancock, who along with others has proposed a quite different location for the lost continent. In his book *Fingerprints of the Gods* (1995) he proposed that the site of Atlantis was in fact Antarctica.

Sticking to Plato's original dates for the existence of Atlantis, he subscribes to the idea that catastrophic displacement of the earth's crust caused the extinction of an advanced civilization that existed on the edge of an extended continent of Antarctica about 11,000 years ago. His contention is that at this time Antarctica was very much larger than it is today and that its most northerly coast reached at least 2,000 miles further north. He further contends that the reason no one has yet located Atlantis is because it lies beneath the frozen wastes of modern-day Antarctica.

In *Fingerprints of the Gods*, Hancock quotes the archaeologists Rose and Rand Flem-Aths (who published their own account of this theory, *When the Sky Fell In*, in the same year as Hancock's book) as saying:

Antarctica is our least understood continent. Most of us assume that the immense island has been ice-bound for millions of years. But new discoveries prove that parts of Antarctica were free of ice thousands of years ago, recent history by the geological clock. The theory of 'earth-crust displacement' explains the mysterious surge and ebb of Antarctica's vast ice sheet.

The link with Plato, they suggest, comes from the idea that some of the records of Atlantis were taken to the area around the Mediterranean by a group of survivors, a group who later seeded the Egyptian civilization, providing them with the technological expertise needed to construct the pyramids, embalm their pharaohs and model the sphinx.

But does Atlantis need to have existed at all? After all, the only account we have to go by is Plato's testament, and although there may have been genuine elaborate legends hidden in lost documents in Egypt (perhaps in the library of Alexandria before its destruction) it is also possible that Plato's story is nothing more than a piece of fiction.

Noting the remarkable similarities between artefacts found in different cultures that developed around the Atlantic Ocean as far apart as Egypt and South America, there are some who hold the view that there had to be a real Atlantis that existed some 11,000 years ago. But the work of such pioneers as Thor Heyerdahl and others has shown that there is no need for recourse to occult explanations for such things; that indeed people travelled more widely at this time than was believed previously.

Nevertheless, the story of Atlantis does lie at the heart of a great occult tradition. The Theosophists (or Theosophical Society), a group that flourished towards the end of the nineteenth century, were particularly enamoured with the legend of Atlantis.

The Theosophical Society was established by Madame Helena Blavatsky in 1875. She wrote several books that have become classics of the alternative historical tradition, including *Isis Unveiled* and *The Secret Doctrine*. The Theosophists believed in what they called the 'Akashic records', what some describe as an 'astral library' – a source of mystical knowledge that may be tapped into by skilled mediums who then divulge secret knowledge to the rest of us. From the Akashic records, Blavatsky and other Theosophists, most notably Rudolph Steiner, constructed an image of a train of seven civilizations or 'root races' (of which we are supposed to be the fifth) dating from the distant past.

According to this chronology, the first root race were invisible, made of 'fire-mist', and lived at the North Pole. The second lived in northern Asia and were almost invisible, but managed to see each other well enough to develop sexual intercourse. The third root race were the Lemurians, who lived in a place called Mu several hundred thousand years ago; the fourth were the people of Atlantis, and we are the fifth. The sixth root race will supposedly return to Lemuria, and after the seventh race has had its time, the human race will leave earth and live on Mercury.

The Atlanteans, who according to this idea are our immediate ancestors (the fourth race of humans), possessed an advanced technology, used flying machines, and had developed sophisticated medical techniques.

It is interesting to note that Theosophists writing in the late nineteenth century were fascinated with the potential of technology, and in some respects their descriptions of ancient lost civilizations bear a marked resemblance to Victorian western culture. For example, the Atlanteans used airships and X-ray machines. This inclination towards imprinting one's own culture on alternative ancient scenarios is exactly what Plato did in his tracts describing the lost continent.

And in more recent times, the myth of Atlantis has found a new impetus amongst believers in the idea that our planet has been visited and perhaps even colonized in the distant past. By amalgamating some of the ideas of the Theosophists and the convoluted hypotheses of such writers as Eric Von Daniken, a large body of people claim to believe that the human race was seeded by aliens, that Atlantis was really the home of an advanced culture that was destroyed perhaps by a nuclear accident or wiped out by an AIDS-type disease. Some are trying to use the Atlantis story as a model for what is wrong with our culture and what could happen if a society becomes corrupt. Strikingly, this is exactly what Plato was doing two and a half millennia ago.

Believers in the idea that there have been advanced human civilizations that have existed and thrived here on Earth in ancient times point to the many and diverse legends incorporating advanced but lost civilizations in our own deep past. And yet, considered objectively, these provide very flimsy evidence for such a bold claim.

The evidence may be broken down into three groups – ancient texts, ancient images and ancient monuments.

The first of these come from a variety of sources and different cultures, including ancient India, China, Egypt and South America. These texts often describe events which could be interpreted (again using contemporary culture as a template) as descriptions of alien visitation, abductions, and even colonization. An example comes from the Old Testament prophet Ezekiel, whose writings have been interpreted by writers such as Eric Von Daniken as coded descriptions of alien visitations, accounts of cosmic travellers who have passed on secret knowledge, perhaps even the original colonizers of earth.

A favourite of the occultists is what has been claimed to be Ezekiel's encounters with an ancient astronaut, taken from the following biblical passage:

Now it came to pass in the thirtieth year, in the fourth month, in the fifth day of the month, as I was among the captives by the river of Chebar, that the heavens were opened . . . And I looked, and behold, a whirlwind came out of the north, a great cloud, and a fire unfolding itself, and a brightness was about it, and out of the midst thereof as the colour of amber, out of the midst of the fire. Also out of the midst thereof came the likeness of four living creatures. And this was their appearance; they had the likeness of a man. And every one had four faces, and every one had four wings. And their feet were straight feet; and the soles of their feet were like the sole of a calf's foot: and they sparkled like the colour of burnished brass.

Upon first reading this might appear to describe something like an advanced flying machine, perhaps one built by aliens, or by the people the Theosophists imagined lived in Atlantis. But it should be recalled that the Old Testament was written by simple people who had little experience of the world, men who lived in constant fear of the forces of nature and the wrath of their God. To them something as natural as a whirlwind or a volcanic eruption could be personified and anthropomorphized into images of strange beings. It is even conceivable that these could be descriptions of real men from a neighbouring, slightly more advanced culture, dazzling simple peasants with chariots, bright ornamentation and well-crafted weapons.

Linked with these texts are records preserved by ancient cultures as verbal accounts. A striking example comes from the West African Dogon tribe, who, according to some accounts, knew of the existence of a star called Sirius B, which can only be seen with the aid of a powerful telescope and was first photographed in 1970. In his book *The Sirius Mystery*, the writer Robert Temple claims that the tribespeople knew this star was a member of what modern astronomers call a binary star system (two stars orbiting each other). Astonishingly, the Dogon tribe even knew the duration of the star's orbit – around fifty years. The Dogon, he claims, learned about Sirius B from the Ancient Egyptians some 3,000 years ago, and others extrapolate further and believe that such astronomical knowledge possessed by the ancient Egyptians was passed on to them by a much older race – once again, the same highly advanced people of Atlantis. However, astronomers are convinced that the knowledge of the Dogon is nothing more than coincidence and point to the fact that a high percentage of star systems are binary and that the figure of fifty years could have been a lucky guess.

The second type of evidence proposed by enthusiasts of these occult ideas is pictorial representation, ancient images that have survived from long-dead civilizations, particularly the ancient Egyptians who, enthusiasts believe, were the custodians of the artefacts surviving the destruction of Atlantis.

Some of these have been widely publicized as proof that we were either visited by advanced extraterrestrials or that there was a race of technologically advanced humans who lived on earth many thousands of years ago. Perhaps the most sensational is a drawing discovered in the ancient Mayan Temple at Palenque in Mexico. It shows a human figure seated in what looks astonishingly like a modern space capsule. The figure is squeezed into a small space jammed with levers and what could be interpreted as control panels. To complete the image there appears to be a plume of smoke emerging from the rear of the contraption, not unlike the vapours that are expelled from a NASA rocket.

This is not the only picture from the ancient world that depicts what could be interpreted as space technology. According to some supporters of the ancient technology theory, primitive man seems to be obsessed with space-suited figures. One drawing found in cave dwellings in Val Camonica, in northern Italy, depicts what may be interpreted as cosmonauts or NASA astronauts. They are dressed in large suits and what look like helmets and visors. Another interpretation may be that the drawings actually show nothing more exotic than the hunting gear worn by primitive people during a period now recognized as a mini ice-age. Similar drawings have been found at ancient American Indian sites in North America, in Uzbekistan and in Tassili in the Sahara.

But for those who want to believe that an ancient people ruled the earth using advanced technology tens of thousands of years ago, the most important link they have to the past is the towering edifice of the Great Pyramid at Giza and the circle of stones at Stonehenge, as well as other sites around the globe.

One of the original Seven Wonders of the Ancient World (and the only one remaining today), the Great Pyramid is without doubt a truly amazing feat of engineering. Known to have been constructed during the third millennium BC, it contains upwards of one million blocks of stone, each weighing about 2.5 tonnes. It measures 230 m (756 ft) on each side and was originally 147 m (482 ft) high – that's four blocks of Fifth Avenue along each side of the base.

Staggering engineering achievement the Great Pyramid may be, but orthodox archaeologists are able to describe in detail how it was built using tens of thousands of slaves dragging the stones from boats which had brought them from quarries in the Lower Nile. They have plotted the route of roads specially designed and constructed to transport the stones, and have shown how Egyptian engineers had the mathematical and engineering skills to construct a building that is not only huge but demonstrates sophisticated number relationships between the length of its sides, its height and the area of the base.

A monument that required comparable engineering genius is Stonehenge. For several decades it has been associated with theories trying to link its construction with alien visitors or ancient humans who possessed technological abilities millennia ahead of their time.

Stonehenge is to be found 13 km north of Salisbury. It was started a few hundred years before the Great Pyramid at Giza, around 2800 BC. But unlike the Great Pyramid, the Stonehenge site evolved over a period of almost 1,800 years. Conventional archaeologists have identified four different phases of construction, with Period I beginning around 2800 BC and period IV ending about 1100 BC.

Theories concerning the use of the site and the way such an edifice could have been constructed by primitive tribespeople are varied and plentiful. Again, enthusiasts of the ancient-astronaut theory suggest that Stonehenge is one of many sites situated on ley lines – hypothetical lines of 'force' or 'natural energy' that intersect at key points.

Although many books, articles and television programmes have been produced debating the idea that Stonehenge is in some way linked with the ancient people of Atlantis, or is perhaps of cosmic significance to extraterrestrials, once more, conventional archaeology can offer a clear picture of how this incredible edifice was built. A growing collection of scholarly works illustrate the many techniques employed by the Ancient Britons to construct the stone circle and show how the materials needed were readily available to its builders at the time.

But whatever the arguments over who built the pyramids and Stonehenge and why, the simple fact remains that any reliance on an occult explanation is at best an insult to human ingenuity. To many (not just empirically minded or sceptical scientists), attempts by occultists to diminish the great achievements of our ancestors is demeaning and crude. But beyond this, the reasoning of people like Von Daniken and other supporters of the idea that the ancients of traditional history could not have done the things they did is simply flabby thinking.

In *Chariots of the Gods?* Von Daniken makes claims such as: 'Is it really a coincidence that the height of the pyramid of Cheops multiplied by a thousand million corresponds approximately to the distance between the Earth and the Sun?'

Well, firstly, the answer is surely 'yes', but, let us give this particular author enough rope to do with as he will. The distance between the earth and the sun is 93,000,000 miles. If we multiply the height of the great pyramid by one thousand million we arrive at a figure of 98,000,000 miles. This approximation is an approximation indeed, out by six per cent; so what does it prove? That the ancient Atlanteans or perhaps visitors from another planet calculate to within a margin of error of six per cent? Nothing else about the great Pyramid is out by even a thousandth of this figure.

On a television programme called 'The Case of the Ancient Astronauts', made in 1978 to debunk Von Daniken, the producers drew an analogy between the author's pronouncements and what a wrongheaded archaeologist of the future might think about our culture. Suppose, they said, in the year 5330 an impressive-looking ancient monument was unearthed in a site known to have been where a civilization once existed, a city thought to be called Washington. Archaeologists of the time calculate that the height of a needle-like construction in the centre of the ruined city when multiplied by forty gives the distance in light years to the second nearest star, Proxima Centauri. Would the obvious conclusion be that the ancient Americans were too stupid to have made this comparison themselves and that the Washington monument was designed and built by ancient alien visitors?

So, where do all these conflicting ideas leave us on the subject of Atlantis? Did it really exist? If it did, what sort of place was it? What sort of society did the Atlanteans have? Was their technology comparable to ours today? Or was Atlantis merely an isolated island kingdom, home to a culture just a little more advanced than its neighbours?

The problem with any suggestion that the Atlanteans had an advanced technology is the matter of what happened to the traces. Hancock's fingerprints of the Gods are the encoded remnants of a great culture he believes can be seen in certain ancient materials and cultural heritage. But would there not be much more that survived? What, by comparison, would our civilization leave behind if it were destroyed? Would people eleven or twelve millennia in the future be able to say with certainty that an advanced civilization once lived on earth?

I think the answer to this is an unequivocal 'yes'. We are a global society and our marks are everywhere. If our culture was to be utterly destroyed, some traces would remain. We have left our stamp in the depths of the oceans, at the highest peaks, on the surface of the Moon, and even in other parts of the Solar System. Atlantis could not have been a truly global civilization, and therefore it could not have been as highly developed as our own culture. In fact it could not have been a society any more developed than that of, say, Europe before the 15th century, when global exploration became common.

And what of the relics, the fingerprints of their time on Earth? For the believer in an alternative history there is no concrete proof, no 10,000-year-old human skull with a silicon implant, no artificial hip joint from 5,000 years before Christ, no laser gun carbon-dated to the time of the ancient Egyptians.

The closest we have come to such a discovery occurred in 1936. Archaeologists working in Iraq stumbled across what is now known as the 'Baghdad battery'. This is a tube a few inches long which consists of all the components of a working cell minus the battery acid itself. It has been dated at about 4000 BC. After making an exact replica of this strange object, researchers used fruit juice as a substitute for battery acid and produced half a volt of electricity.

Many decades after it was first discovered, the Baghdad battery is still believed to be genuine, and no one has been able to explain its origin. It may be a relic from an ancient technological society, but it is strange that such a find is entirely isolated. Advances in technology are inter-linked. It is extremely unlikely that a car, for example, could be built unless the society in which its inventor lived had access to techniques for producing metal sheets, petrol or some other fuel, materials for the tyres, and the machine tools to build the individual components.

It might be that the Baghdad battery was constructed by an unknown genius, a Leonardo da Vinci of his time who stumbled across the secret of electricity and built it from scratch. But this is nothing more than a fascinating possibility.

Apart from the marked lack of evidence, there is nothing intrinsically wrong or contradictory about the idea that a reasonably advanced civilization could have sprung up and flourished for a while on Earth, perhaps tens of thousands of years ago. Indeed there are no real intellectual objections to the idea that we could have been visited by aliens in the dim and distant past and that an advanced culture was seeded by colonizers or a small group stranded here. But equally, as I have said, it is a quite unnecessary hypothesis.

As for the location of the lost continent, the smart money is still with the idea that the Minoan civilization was the origin of the Atlantis myths, which were handed on to the ancient Egyptians who wrote the earliest account. This could have been the basis for Plato's story, a ripping yarn that has been elaborated further still in recent centuries, and a tale that will almost certainly continue to be elaborated upon in the centuries to come.

7

From Here to There and Back Again

Is Teleportation Possible?

For my part, I travel not to go anywhere, but to go. I travel for travel's sake. The great affair is to move.

Robert Louis Stevenson

Strictly speaking, the Tardis does not teleport except on the rare occasions it moves through space while remaining in the same time. But the Doctor as well as his adversaries have sometimes used various methods to teleport themselves. Indeed, the Doctor's most ruthless enemy, the Master, expended great efforts in developing a method of teleportation at Cambridge University, where he disguised himself as Professor Thascales. The Daleks and other, lesser, enemies, including the War Lords and the Nimon, have all employed different methods of teleportation, while some of the strangest creatures to appear in the programme have displayed a natural ability to teleport themselves from one location to another. But exactly how could this be done, if at all? Is teleportation merely a flight of fantasy that can never be realized, or is it possible that an advanced civilization such as the Time Lords could work within the laws of physics to construct a machine that could teleport solid objects?

The idea of teleportation was actually popularized in the late 1960s by the TV show *Star Trek*. The transporter was one of the mainstays of the programme and was put into the original show for the prosaic reason that the cost of landing a huge spacecraft, the USS *Enterprise*, was considered too great a drain on the special-effects budget. To circumvent this problem, the creator of the show, Gene Roddenberry, came up with the ingenious idea of simply keeping the *Enterprise* in space and transporting the crew to the surface of an alien world whenever the need arose.

But the concept of teleportation goes back much further. It was employed by the science-fiction writers of the 1940s, and in the occult tradition it is considered a magical power possessed by certain spiritually advanced beings: the most enlightened Tibetan lamas, for example, are said to be capable of teleportation.

However, for all the ingenious uses to which teleportation has been put in science fiction and for all the faith that devotees of the occult have placed in the concept of teleporting material objects, using what we know of physical laws it would be rather difficult. Indeed, it is no exaggeration to say that the idea of teleportation is one of the most far-fetched in all science fiction.

There are really two types of teleportation. The first involves the process of transferring real matter instantaneously from point A to point B. The second is the transfer of information. This, it seems, might be a rather less demanding alternative.

First, let's consider the option of matter transference, actually taking apart an object and moving its constituent particles to another location. Would this ever be feasible?

The simple answer to this is almost certainly 'no'. And once again, just as we saw with time machines and interstellar travel, one of the most stubborn obstacles to success is that the process would require almost unimaginable amounts of energy. But why should this be so?

The first thing to consider is the level of detail we need to work with – how far do we need to decompose an object to make it teleportable? Would it be sufficient to deconstruct the object to the atomic level? Maybe, but the problem with this is that atoms are not the most fundamental objects in the universe. They are made of smaller sub-atomic particles, protons, neutrons and electrons. And these are in turn comprised of still smaller entities called quarks, which, as far as we know, are the most fundamental constituents of matter.

As we go deeper into the sub-atomic world the energies needed to separate out the parts increase exponentially. What is called the 'binding energy' of particles is much greater between quarks than it is between the constituents of the atomic nucleus, the protons and the neutrons, and the binding energy between these units is far greater than that between atoms themselves.

So, for the sake of simplicity let's imagine that our teleportation device works on the principle of breaking an object into just its constituent atoms.

I say 'just', but this is where we hit our first major difficulty. It has been estimated that the human body contains around 10^{28} atoms. To put this into perspective, a recent estimate of the number of grains of sand in the world (based on the premise that the sand is spread out to produce a layer 10 cm thick, over the entire surface of the earth) came up with the figure 10^{22}. In other words, the number of atoms in a human body is approximately equal to all the grains of sand on one million Earths.

To break these atoms apart, work has to be done to overcome the binding energy. And to break a human being into atoms turns out to require a rather large amount of energy. If we say the average man weighs about 85 kilos, by putting this figure into the equation $E = mc^2$, we get a figure for the energy equivalent of a man as being equal to almost 2,000 megatonnes of TNT, or about 200,000 times the energy released by the atom bomb dropped on Hiroshima.

So, even if we ignore the technical difficulties of teleporting 10^{28} atoms across space and repositioning them precisely as they were in the original, the energy needed to create this collection of atoms in the first place is massive.

In order to address the matter of repositioning this vast number of atoms, it would be best if we leave matter transference and explore the possibilities of what some have considered to be a slightly less daunting endeavour – the teleportation of information. This would allow us to replicate an object at a distance, a process some experts refer to as 'biodigital cloning'.

Some forms of 'teleporting' information are already familiar to us, but I put this word in quotation marks for a good reason. The telephone, the fax machine, the internet and, indeed, the book are, if you like, teleportation devices. They are each vehicles for the passing on of information. The difference between these things and a real teleporter of information is that the latter would create a perfect replica of the original collection of information, in three dimensions and instantaneously. What this really means is that an information teleporter would convert a physical object into its information content, transport this information and from this reconstruct the original precisely at a different location. This would inevitably destroy the original.

But just how easy is this to do? Well, as you might expect, there are a number of thorny problems attached to the idea. To begin with, the information content of even a simple physical object is huge. The location of each constituent atom, its own characteristics and its relationship with all the other atoms in the object amounts to a great deal of information.

Suppose we wanted to teleport the information content of a human being. First we would need to collect the information. Next, this information would need to be transported from point A to point B. Then, at its destination, the information would have to be used to reconstruct an exact replica of the traveller.

The first problem to face with this scenario is the sheer quantity of information involved.

Again, I will assume that our teleporter works on the atomic level (rather than sub-atomic particles or quarks). If we need to extract the information describing the atoms of the traveller, we will need to know something about each of the atoms, its position relative to the other atoms, its internal characteristics, the bonds each atom forms with other atoms, and much more. Remember, we are trying to transport everything we need to know about a person so that we may create a perfect replica at the other end of the teleportation process.

When we talk about information, we deal in 'bits' and 'bytes'. A bit is a 'yes' or a 'no', an 'on' or an 'off' – the most fundamental piece of information. In IT jargon, these are gathered together in bundles of eight which are then called 'bytes'. If we work on the principle that each atom in the human body requires, say, 1000 bytes (a kilobyte) of information to define it, then the information content of a human consisting of 10^{28} atoms would be 10^{28} kilobytes.

To give this number some meaning, consider a storage disk such as the portable USB drive I'm using on my computer. This has a capacity of 64 megabytes (64 Mb) or 64 million bytes. So, to store the total information content of a human being on a set of these USB drives, I would need approximately 1.5×10^{21} of them. Returning to my earlier analogy of grains of sand, this would be about one sixth of the total number of grains of sand in the world.

But for the moment, let us imagine we can store this information. Storing is one thing, but we then need to be able to interpret the information and to reconstruct the original *exactly* as it was. In other words, we need a device that can process this amount of information extremely quickly. The fastest information retrieval systems in use today work at a rate of about 1,000 megabytes per second. This means that using present-day technology, to extract 10^{28} kilobytes of information from a storage system would require 10^{22} seconds. This is about one hundred and fifty times the age of the universe.

Looking at these facts, it would seem that even the teleportation of information content, a system that would allow us to reconstruct a human, is an impossibly difficult thing. But, surprisingly perhaps, there is a glimmer of hope.

In 1965, during the early days of commercial computing, one of the pioneers in the field, a co-founder of Fairchild Semiconductors, Gordon Moore, made the observation that computer advance was exponential and predicted that computing power would double every eighteen months. The press dubbed this pronouncement 'Moore's Law', and it has been shown to be remarkably accurate for more than four decades.

It is easy to see how accurate Moore's prediction has proven to be. When I purchased my first computer in 1988 (a trusty Amstrad), it contained about one million microprocessors. The Mac I'm working on today has a capacity a thousand times greater. If we extrapolate into the future using Moore's Law, we reach the surprising conclusion that 200 years from now computers will, in theory, be able to handle the sort of information content associated with the teleportation of a human being.

As I've mentioned before, it is quite possible that alien civilizations could be thousands of years more advanced than us. In which case a matter of a mere two centuries is inconsequential. The computer systems used by an advanced culture would have no problem managing information contents of 10^{28} kilobytes. But are there other, more fundamental, problems with the idea of teleportation?

One aspect of the matter which is worth mentioning just in passing (because it is beyond the scope of this book) is the question of spirituality and religion. In the above scenario, we are considering the possibility of breaking a human being into their most fundamental parts, in effect obliterating them. We then propose to rebuild them, atom by atom. What then of the concept of the soul or the spirit? Is a human being nothing more than the sum of their atoms?

As an atheist, I have no problem with this, but there are many who would disagree. In effect, the process of teleportation involves death and rebirth. This may cause ethical dilemmas for advanced civilizations, even if the technical hurdles could be overcome.

Let us assume, though, that an advanced civilization has no ethical objections to teleportation and they have the computing power to facilitate such a process. There remains one other problem to solve. This is the fact that Heisenberg's Uncertainty Principle prohibits atomic analysis in the way we would need in order to build a teleportation device.

I have mentioned this principle in earlier chapters. It is one of the fundamental tenets of quantum mechanics, and when trying to conceive of ways to travel in time or to employ wormholes, it proves to be as much a party-pooper as the condition of the universal speed limit, c.

In 1927 Werner Heisenberg suggested the idea that at the heart of the atom and all things in all universes lies *uncertainty*. It underpins everything. Heisenberg realized something about the sub-atomic world that had not occurred to anyone before, which is that it will always be impossible to know precisely the nature of sub-atomic particles. If, for example, we know the position of an electron, we cannot at the same time also know its momentum. If we know its momentum, we cannot know exactly where it is. The same is true for any pair of properties displayed by particles that are approximately the size of an electron.

But why should this be so?

The answer comes from the fact that in order to observe anything we have to use radiation. Radiation is a wave or a particle (depending upon the way you interpret it), but either way, each wave or each particle of radiation is about the size of an electron (or its associated wave) so when the radiation is bounced off the electron we wish to observe, the position of this electron is changed – it is knocked off course.

There are no 'special forms of radiation'. All radiation has a wavelength, or if you want to think of radiation as a stream of particles, those particles have a momentum, they have energy. They will interact or interfere with the very small thing you are observing, such as an electron.

Of course, this phenomenon operates in the everyday world of larger objects, but because the particles or waves of light are so very much smaller than a massive object like a train or a car, or you and me, we do not notice the effect.

Now, you may be wondering why it should matter whether or not we can know simultaneously two variable properties of a particle the size of an electron. But actually this fact has staggering implications, because it means that the universe is built upon uncertainty. It means if we cannot know about the properties of the sub-atomic world with absolute certainty, then we cannot know about *anything* with certainty.

What this means for the designers of a teleportation device is that it is not possible to collect all the information needed about an object (such as a human) before transporting it and reconstructing the original. As soon as we know the momentum of an atom, we no longer know its position. The same goes for any paired pieces of information in the atomic and sub-atomic world.

For some time this has led scientists to the conclusion that teleportation would always be an impossibility no matter how much technology was to advance. Heisenberg's Uncertainty Principle is a universal law and, like the limitation to sub-lightspeeds, it is an unbreakable one.

But just as scientists have explored theoretical ways in which Einstein's speed limit could be bypassed, new concepts are emerging which show that there may be a way to use quantum mechanics to overcome the problem of uncertainty. And the source of this possible get-out clause comes from a proposition made by Albert Einstein and two colleagues, Boris Podolsky and Nathan Rosen, in 1935, soon after Heisenberg proposed his principle.

Einstein was always suspicious of some of the more esoteric aspects of quantum theory emerging during the late 1920s, a stance that led him to make his famous remark, 'God does not play dice.' Because of this he was tireless in his efforts to try to show that Heisenberg was wrong about his uncertainty principle, and he spent years coming up with one idea after another that he believed would dispense with the uncertainty of the quantum world. However, every idea he came up with was knocked back by Heisenberg, Niels Bohr and the other quantum evangelists of the time.

Then, while he was working at the Institute of Advanced Study in Princeton, Einstein and two colleagues developed what they thought would be a foolproof argument against quantum uncertainty. It became known as the EPR Paradox after the three physicists who developed it, and it goes something like this. Einstein and his colleagues proposed a thought experiment involving two particles, let us say, for the sake of argument, two electrons. Now remember Heisenberg's Uncertainty Principle – that we cannot know the precise nature of the sub-atomic world because we cannot know exactly any pair of variable properties of particles. But, Einstein said, imagine we know the precise position of the two particles at the start of the experiment and that they are very close together. Within the accepted rules of the quantum theory, we can measure the *total* momentum of the two particles, it is just that we cannot say what the momentum of each would be.

Now picture a box. In this box, we have two electrons (electrons A and B) that are a known distance apart and we know their *total* momentum. Next, we must imagine sending these two particles in opposite directions at the speed of light through coiled tubes at each end of the box.

Einstein now asserts that if we measure the momentum of one of these electrons we automatically know what the momentum of the other one must be, because we knew what the total momentum was in the beginning and this total cannot have changed. Then we can work out the position of the first electron. This will of course change the momentum of the particle – remember Heisenberg: the very act of observing will alter the properties of the particle being observed. But, and this is the crucial point, it will not, Einstein asserted, alter the momentum of the other particle. We can measure the position of the second particle and we know its momentum because we have calculated the momentum of the first one. Electron A goes off along the tube to, let us say, the left of the box. Electron B shoots off at the speed of light into the tube on the right. Sometime later, we measure the momentum of the first particle in the right-hand tube. We knew the total momentum, so we now also know the momentum of the electron in the left-hand tube. We can then try to measure the position of the particle, electron B, in the right-hand tube.

This measurement will be messed up because, by measuring it, we disturb it, so we do not really know its position. But we have not measured the momentum of the particle in the left-hand tube. We just deduced it by knowing the total momentum and the momentum of the particle in the right-hand tube. So we can measure the position of the particle here in the left-hand tube, because we can know one property without disturbing the particle.

Einstein claimed that the only way that this arrangement could fail is if the particles could somehow communicate over a distance and disrupt the process. Niels Bohr and Heisenberg agreed and counter-claimed that if this experiment was actually performed something must happen that does mess up Einstein's plan. They said that, some-how, if you alter the momentum of the particle in the right-hand tube, electron B, this will also instantaneously alter the momentum of the one in the left-hand tube, electron A.

Einstein was convinced this was impossible. He had laid down one of Nature's universal rules – that nothing could travel faster than light. These particles were moving apart at the speed of light; how then could one communicate with the other across space? How could the change in momentum of the particle in one tube alter the momen-tum of the other electron in the other tube instantaneously? He utterly refused to believe it, but Heisenberg and Bohr stuck to their guns.

From this argument emerged the term 'entanglement', which physi-cists use today to describe the manner in which two or more particles can communicate in the way Heisenberg and Bohr believed. Einstein referred to it as 'spooky action at a distance'. But although no one really understands how it happens, it is an effect that is being used increasingly by those who study quantum properties.

It was not until the 1980s, some thirty years after Einstein had died, that someone actually conducted an experiment similar to the one Einstein, Heisenberg and Bohr had discussed. And what was the outcome of this experiment? It showed that Einstein had indeed been wrong. Known as Bell's experiment because it was based upon a thought experiment devised by the physicist John Bell, it has subsequently become one of the most famous and puzzling experiments in the history of science.

Bell's concept had its origins in the thought experiment of Einstein, Podolsky and Rosen that had brought forth the EPR Paradox. Imagine two particles which are interacting in some way – maybe they have been fired from a shared source, a particle generator or a cyclotron. Now, imagine these two particles are fired into a device that sends them off in opposite directions at the speed of light. Next, Bell imagined altering a property of one of the particles, say the one travelling left. He altered the spin of the particle, perhaps, or the direction in which it vibrated as it continued on its journey.

Bell began to wonder what, if anything, would happen to the other particle as a result of making this change to the first one, and according to his calculations, the second particle *was* affected by the change made to the first one. But that was not all. It was affected *instantaneously*. In other words, the first particle had somehow 'told' the other particle that it had experienced a property alteration *the very instant it happened*.

But the consequences of this must mean that the two particles were communicating faster than the speed of light, transgressing one of the central tenets of physics – that the speed of light is the upper limit in our universe. Worse still for those who doubted the validity of quantum mechanics, the actual experiments conducted two decades after Bell's thought experiment confirmed his conclusion. And since then research groups all over the world have repeated the experiment thousands of times, and they always get the same result: the particles seem to have the ability to communicate with each other faster than light.

This weird property, the 'entanglement' that Einstein, Heisenberg and Bohr had argued over, lies at the heart of new research to create a form of teleportation. I say 'a form of teleportation' because, so far, experiments have been limited to transferring information about what are called 'quantum states', that is, properties of sub-atomic particles.

In 1993 a team of scientists working at IBM led by Charles Bennett described a way in which entanglement could be used to teleport information. Publishing their findings in the 29 March 1993 issue of the respected journal *Physical Review Letters*, the team explained how it might be possible to teleport the information content of a single photon of light (and, in principle, the simplest sub-atomic particles).

Five years later a team at Caltech brought this idea to reality by teleporting a photon a distance of just over a metre. Then in 2004, two teams, one in the US and the other at the University of Innsbruck in Austria teleported information contained in a charged atom. They did this by manipulating a pair of entangled, charged atoms (or ions), that is, two particles that share quantum information as those in Bell's experiment could. These ions are called B and C. Next, the state or characteristic to be teleported is created in a third ion, A. Then, one ion from the entangled pair – say B – is entangled with A. The internal state of both of these is then measured and the result sent to ion C. This then transforms the quantum state of ion C into the one that had been established for A (the information we wanted to teleport). In conducting this transfer, the original quantum state of A is destroyed.

This is, of course, a very long way from developing a technique to teleport a person. As we have seen, the information content of a human is in the order of 10^{28} kilobytes – very much larger than the amount of information dealt with in these experiments. But what they show is that the laws of quantum mechanics can be stretched to allow for the principle of teleportation, and the researchers have succeeded in taking the concept from pure mathematics to a real laboratory situation. The hope is that these experiments will be refined and that gradually larger amounts of more complex information may be manipulated.

Naturally, these ideas have generated great excitement, not just because of the fascinating theoretical consequences and the weirdness of such work; these experiments could soon lead to very real and useful applications. It is hoped that in the not-too-distant future the idea of quantum teleportation will be used to develop what is called a 'quantum computer', a device that would be enormously more powerful than a conventional computer.

So much then for the idea of teleporting information. The technology required to transport the information content of a human lies far beyond anything we can even imagine today, but the principle does not, it would seem, transgress the laws of physics. Furthermore, if we are to consider a culture that is more advanced than ours by many thousands of years (such as the civilization of the Time Lords or the Daleks) then we should not dismiss the idea that their computing systems might be capable of handling the huge amounts of information that constitute a living being. For now, though, as with journeys though hyperspace and time travel, the notion of building a teleportation device goes no further than the pages of science-fiction novels and the scripts of *Star Trek* and *Doctor Who*.

8

Robots and Mechanoids

How Clever Will Computers Get?

> *Science fiction writers foresee the inevitable, and although problems and catastrophes may be inevitable, solutions are not.*
> Isaac Asimov

As a child I always found the Cybermen the most frightening of all the Doctor's adversaries. I think it had something to do with their cold, expressionless faces, and the utterly clinical way they went about things.

Strictly speaking, Cybermen were not robots or androids, but cyborgs, an alien intelligence that had adapted itself using cybernetic implants and organs. But the Doctor has faced some pretty unpleasant robotic enemies, including BioMechanoid, a cyberdragon from the planet Svartos, the Yeti, and of course the very unpleasant Robots of Death.

Like time travel, interstellar voyages and teleportation, robots are a staple of traditional science fiction dating back to Mary Shelley's *Frankenstein*, written in the early 19th century. The most famous robot stories of the 20th century came from the pen of Isaac Asimov, who wrote scores of short stories including 'I, Robot' and 'The Rest of the Robots', along with a handful of acclaimed novels in which robots and androids played key roles.

These stories are particularly interesting because Asimov was a scientist and understood that if robots were ever to become a reality they would have to be programmed to work with humans. This led him to create his famous Laws of Robotics, which he defined as:

Zeroth Law: A robot may not injure humanity, or, through inaction, allow humanity to come to harm.*

First Law: A robot may not injure a human being, or, through inaction, allow a human being to come to harm, unless this would violate the Zeroth Law of Robotics.

Second Law: A robot must obey orders given it by human beings, except where such orders would conflict with the Zeroth or First Law.

Third Law: A robot must protect its own existence as long as such protection does not conflict with the Zeroth, First, or Second Law.

One of the most striking things about Asimov's classic stories from the 1940s and 1950s is that he never once mentions the computer, but robots abound. The reason for this is almost certainly the influence of a much older classic than 'I, Robot', Shelley's *Frankenstein*, which implanted the idea of the automaton or artificial being into Western literature.

Beyond the world of science fiction, though, the manufacture of an efficient robot similar to the ones Asimov wrote about, or with the power of the Robots of Death, could certainly not be possible without major advances in computer technology. The production of a sophisticated, humanoid robot that can obey spoken commands and with which we can communicate on an intelligent level, is still several decades away. But robotics is a growing field and one that is attracting some of the most imaginative researchers as well as considerable financial backing.

The Chinese were the first people interested in calculating numbers using a machine, and they invented the abacus about 5,000 years ago. Between this time and the start of the modern scientific age, many thinkers, philosophers and engineers devoted entire careers to developing labour-saving devices to make calculations faster and more accurate.

---

* Asimov created the First, Second and Third Laws in 1940 for his first robot stories. The Zeroth Law came much later, in 1985, when it was created for a particular plotline. Because it was more fundamental than even the First Law, Asimov dubbed this law the Zeroth Law.

Renaissance mathematicians were fascinated with the idea of developing some form of mechanical calculator that would be a great improvement on the traditional abacus. One of the most far-sighted innovators of the time was the great 16th-century mathematician Luca Pacioli, who is sometimes referred to as 'the father of accounting'. He designed a primitive calculating machine for speeding up complex financial transactions, and his friend Leonardo da Vinci also tried to devise machines that could calculate faster than humans.

During the 17th century, the world was graced with a collection of remarkable mathematicians. Newton is perhaps the most famous, but his great rival, Gottfried Leibniz, was every bit as gifted, and he was far better at communicating his ideas. Leibniz once declared: 'It is unworthy of excellent men to lose hours like slaves in the labour of calculations which could safely be relegated to anyone else if machines were used.' He then went on to create the binary system of calculation which lies at the heart of modern computing. A few years later he designed and built a mechanical calculating machine that became the talk of the day within scientific circles while greatly boosting his reputation throughout Europe.

One of the most famous early 'computers' was designed by Charles Babbage, who held the Chair of Mathematics at Cambridge University, the Lucasian Professorship, once occupied by Isaac Newton and today held by Stephen Hawking. Babbage was obsessed with calculating machines, and he spent a large portion of his career perfecting a device he called a 'difference engine'. A wealthy man, Babbage spent £6,000 of his own money in building his device and secured a further £17,000 from the British government after persuading them of the importance of his invention.

Sadly, though, Babbage's machine never performed as well as its creator imagined it could. Like others excited by the idea of calculating machines, Babbage was too far ahead of his time. He was attempting to build a calculating machine during the 1820s, a mere generation after the first battery had been created by Alessandro Volta, during a time in which the very idea of electricity lay at the fringes of intellectual investigation. Babbage's machine relied upon a vast number of mechanical switches working in unison, and each part had to be handcrafted with incredible precision. But without electricity this device could never have worked in the way a modern computer does.

The first true computers did not appear until the Second World War. One of the first was a machine called ENIAC (Electronic Numerical Integrator and Computer). It was built at the University of Pennsylvania in 1943 by John Eckert and John Mauchly and weighed thirty tonnes, guzzled electricity, took up 1,800 square feet of floor space and possessed a computing power roughly equivalent to the rhyme-playing chip in a novelty birthday card.

A huge amount was learned from the earliest computers employed during the closing years of the Second World War, and because of parallel advances in electronics, which had also been spurred on by the military, in the early post-war years the entire science of computing began to advance more rapidly than might have been expected.

Key to the advances in computation speed was the arrival of more and more sophisticated circuitry. The early electronic computers used vacuum tubes, but these are cumbersome and fragile things, and the first ENIAC used no fewer than 18,000 of them. The evolution from ENIAC to the latest generation of computers we use today happened thanks to a string of developments in computing, including the creation of programming languages, improved methods of inputting and outputting information, the development of a vast range of software, and the enormous and rapid progress in electronics needed to produce increasingly sophisticated hardware.*

The transistor, which replaced the vacuum tube, was developed soon after the Second World War by a team at Bell Laboratories. The transistor was much smaller and cheaper to make, and it revolutionized electronics by making it possible to manufacture circuit boards that were tiny compared with the vast array of tubes and wires that went before them.

But the transistor was merely a stepping stone along the path of designing faster and smaller computers. The microchip arrived in

---

* For me, it is a sobering thought that when I started university in the late 1970s I had to use stacks of punch cards to input data and make simple calculations on the 'college computer', a vast machine that was kept in the basement of the main building. A technician would take my cards and I had to wait until the following day for the answers and a print-out.

1961, and it marked another massive advance. Instead of a circuit with large components linked by wires, the microchip works by employing eight different layers of semi-conducting silicon and insulators, with the components and the wiring embedded in the layers. This means that elaborate circuits can be made much smaller.

Two of the earliest big projects to utilize the microchip were the American and Soviet space programmes. Apollo 11, the vehicle that first took men to the Moon, contained no fewer than one million integrated circuits, most of which were experimental microchips in the computer systems of the spacecraft. The knowledge acquired from using this embryonic technology spawned today's computer and internet industries, which together now earn approximately one trillion dollars a year and employ an estimated six million people in the United States alone. But aside from this boon, the development of the microchip, accelerated by the space programme, has altered society profoundly.

Since the first microprocessors appeared, manufacturing techniques have improved enormously and chips are becoming smaller and more powerful with each new generation of design. As I mentioned in the last chapter, this advance was put into context very early in the evolution of computers when Gordon Moore made the observation that computer advance was exponential and that computer power would double approximately every eighteen months.

When Moore made this declaration, a computer used in a large corporation contained fewer than 100 transistors; today the average home pc contains in excess of 1,000 million (1,000 times more powerful than the computer systems on Apollo 11). Furthermore, this trend is set to continue until the size of chips becomes constrained by the limits of the physical world, a trend that will only be halted by one of those troublesome universal laws that have cropped up in almost every chapter in this book.

When designers reach the point where they are able to shrink chips to the size of molecules, these processors will become subject to Heisenberg's Uncertainty Principle, and this condition will actually prevent them from functioning as desired. If Moore's Law continues to hold true, this limit is expected to be reached around the year 2020. One possible way around this problem is the quantum computer (mentioned in Chapter 7). This works on the principle of teleporting simple units of information between sub-atomic particles, and such a machine would not be limited by the uncertainty principle.

Advances in the design of computers is obviously a key element in the future evolution of robots. If designers can cram more processing power into a smaller space, it means that more complex and capable artificial brains can be produced. The human brain is a masterpiece of bioengineering, the most sophisticated machine we know of and one that has evolved over millions of years. In order to produce a robot with which we can communicate via speech, a machine that can follow complex instructions and shows the same sort of versatility and subtle behaviour as those described in science fiction, it will need to have a very powerful 'brain'. At the same time the robot needs to be mobile and of a comparable size to a human. All of which means that a great deal of computing power will have to be squeezed into a head not much bigger than that of the average human.

In order to make this happen, we not only have to employ incredibly advanced computers – indeed, it is likely that only a quantum computer would do – we will need to become masters of another key discipline, called 'nanotechnology'. This is the relatively new science of the very small, where tiny machines are controlled either by a central master computer or act autonomously with pre-programmed instructions.

The name 'nanotechnology' comes from the unit of measurement the 'nanometre', or one billionth of a metre, because this is the size-scale in which nano-machines operate. To give some idea of just how incredibly small this is, a human hair is about 100,000 nanometres in width, and a microchip that can fit on the tip of your finger is about five million nanometres across.

So, how can machines this small be made and controlled by humans? The answer lies in the way machines are built. Today, engineers construct machines from *the top down*. In other words, they try to miniaturize them further and further. This is a trend we have seen in the microelectronics industry during the past three decades. But there is a limit to how far machines can be made smaller by simply miniaturizing their parts, and many engineers believe we have almost reached this limit. The alternative way of building a tiny machine is from *the bottom up*; constructing them molecule by molecule.

This was first suggested by the great American physicist Richard Feynman, who in 1959 declared that 'There is more room at the bottom.' At that time, the computing power of a modern mobile phone was available in only a few universities in the world and required a machine the size of a school hall. As we have seen from Moore's Law, with processing power increasing a thousand-fold each decade, memory capacity has had to keep up, which means that the devices that store each unit or 'bit' of memory have had to become smaller and smaller. By extrapolating this progress it is apparent that by about 2020 quantum computers will allow the development of a machine the size of a virus but possessing the computing power of an Apple notebook. Give these devices mobility and communication links with a master computer controlled by humans, and you have workable nanotechnology. These nano-machines could then build an object from the bottom up, each tiny robot working in unison with billions of others.

This might sound far-fetched, but it is exactly what Nature does all the time. A flower grows from a tiny collection of cells, which divide and produce new cells; such mechanisms create the breath-taking diversity of life we see all around us. Bacteria are really nothing more than tiny robots that perform specific, simple tasks. They are propelled by tiny flagella (rapidly moving 'oars') and driven by a tiny biochemical 'motor'.

At the moment, scientists are learning how to arrange individual atoms and molecules and to build simple structures with them. In 1990, scientists at IBM wrote the company name using individual atoms, stacking them six high. To do this they used a newly developed machine called a 'scanning tunnelling microscope' (STM) which allows them to 'see' individual atoms using a complex computer enhancement technique.*

---

* It's important to note here that moving atoms into position or 'observing them' with an STM does not transgress Heisenberg's Uncertainty Principle. This rule only comes into effect if we wish to know more than one of a pair of quantum character-istics; moving atoms is allowed as long as we do not also wish to know their momentum.

The potential applications of nanotechnology are almost limitless, and we are only now realizing ways in which nano-machines might be used. It really is a science bounded only by the limits of our imagination. One of the most obvious applications of 'nanobots' (or 'nanites' as they are sometimes called) is to use them in the operating theatre.

Imagine a surgical procedure a few decades from now. A woman is diagnosed with breast cancer. Instead of having to suffer the rigours of radical surgery or chemotherapy, she undergoes one simple two-hour procedure and the cancer is removed. The surgeon inspects the tumour, and instead of wielding his scalpel he instructs a computer operator to call up a prearranged programme and injects a specially prepared solution into the tumour site. The syringe contains several million nanobots, each so small that thousands of them could fit onto a pinhead. The computer operator then directs them to the tumour site and they begin their work of eradicating the tumour atom by atom, resealing blood vessels and applying tiny quantities of prepared drugs to the appropriate areas. By the end of the procedure all trace of the tumour has been removed without the need for any form of invasive surgery.

This technology could of course be applied to a raft of other medical procedures. Special bots could be sent into the body to reconstruct the damaged limbs and organs of accident victims, to deliver special drugs directly to relevant cells, to stimulate the regrowth of mal-formed bones or to clear blocked arteries.

Building a robot that can repair damaged tissue is, though, a long way from spelling IBM with individual atoms, but nanotechnology is evolving at a frantic pace. The first stage of this evolution is to 'see' what you are dealing with and finding the skills to manipulate individual atoms and molecules. The next is to build simple devices and to then gradually make these more elaborate and versatile.

The world's first artificially constructed molecule has been made by bringing together one atom of oxygen and one of carbon to make carbon monoxide; and in recent years cutting-edge nano-engineers have managed to construct tiny rotors about the size of a spec of dust. These are small enough to travel along a blood vessel. The next step will be to design machines that can be controlled en masse by an external computer and able to carry out specific functions rapidly.

Many scientists believe that within twenty-five years nanotechnology will be the most important technology in the world; and ultimately, scientists and engineers hope to construct cars, houses, spacecraft, almost anything, from the bottom up, 'growing' objects in vats using billions of tiny 'assembler' and 'replicator' nanites. This, after all, is the way a foetus is grown in the womb using DNA and RNA, Nature's computer programmes and templates.

Given the vast potential of nanotechnology, it is not surprising that a good proportion of research funding is coming from the military. Defence research organizations including the MOD's Defence Evaluation and Research Agency (DERA) and the American equivalent, the Defense Advanced Research Projects Agency (DARPA) have each been spending hundreds of millions of dollars annually on developing military applications for nanotechnology. Currently on the drawing-board and not far from use in the field are insect-sized spybots that can fly into an enemy zone and remain invisible to radar or human observers; 'killerbots' that can be targeted with incredible precision; and even 'nano-dissemblers' that can be implanted and instructed to deconstruct enemy facilities molecule by molecule.

For many people, these applications are a cause for concern. Some analysts envisage a future in which wars are fought by robots (both nanobots or man-sized machines); and, although this may reduce human casualties, there are also dangers associated with these advances.

The most important of these is the fact that just as nanites may be used to build things from the bottom up, they can also be employed to deconstruct anything molecule by molecule. The upside of this is that these machines can be used to decompose waste very efficiently. The downside is that if self-replicating nano-dissemblers were let loose deliberately or slipped out of control of the military they could wreak havoc. This scary possibility has been dubbed the 'Grey Goo Scenario' because if something went wrong with the systems which regulate the self-replication of nanobots they could begin to reproduce like a virus. In the worst-case scenario they would be virtually unstoppable and break everything down to grey goo.

It is obvious then that nanotechnology offers huge potential, but this new science is also crucial to any development in the abilities of android-type robots, because the technology required to pack more processing power into a restricted space in order to build an android is the same as that now being refined to construct nano-machines.

To give some idea of the level of computing power needed to build a robot with the same brain capacity as a human, let's make a rough comparison between a specific brain function and the computer processing power this would require. A good example is the retina of the human eye, a patch of receptors and image-processing circuitry which is about a millimetre thick and two centimetres in diameter.

The part of the retina containing the image-processing 'hardware' is actually only one tenth of a millimetre thick, but it is able to detect boundaries between dark and light and movement in approximately one million image regions simultaneously. Information from each of these regions is detected in the retina and communicated to the brain via the optic nerve, and it can perform around ten individual observations per second.

Computer scientists using artificial eyes have found that a machine able to 'see' each region of movement in the visual field requires about 100 computer instructions. So, to replicate the function of a human retina would need ten million times more computer instructions, giving us a total of 1,000 million instructions per second, or 1,000 MIPS.

Now, if we scale up from the retina to the entire human brain, we begin to get a picture of the vast computing power inside our heads and the enormity of the task in replicating this in an android. If we say the image-processing equipment in the retina weighs about 0.02 grams, a typical 1.5 kilogram human brain would need 75,000 times more processing power than that used by the circuitry in the retina. Rounding up the numbers this comes to about 100 million MIPS, or 100 trillion instructions per second.

These calculations are very approximate, but they offer some interesting insights. Current computing power stands at around 1,000 MIPS, just enough to process the information derived from a retina alone. But, as we have seen before, processing power is doubling every eighteen months to two years. At this rate, the processing power of a computer will rise from the current 1,000 MIPS to the 100 million MIPS required to replicate the processing abilities of a human within about forty years.

Robot designers talk about the possible future generations of robots. The experimental machines of today are considered first-generation machines which use processors that can carry out about 1,000 million instructions per second and allow robots to move, to 'see' to a very limited extent and to 'learn' in the most rudimentary sense – they can respond to external stimuli and develop responses or learn to find faster ways to perform a rudimentary task.

Within a decade, advanced first-generation robots will have 'brains' with a processing capacity of 5,000 MIPS. This is equivalent to the mental abilities of a lizard. With these capabilities they could perform the most basic household duties following a pre-programmed set of instructions. Yet some experts say that getting a robot to perform even simple chores is not as easy as it sounds. For example, a robot with the sort of mental capacity of 5,000 MIPS would be incapable of distinguishing between a pile of washing and a sleeping cat.

Second-generation robots would have the processing power of a mouse (about 100,000 MIPS). They would be the first robots to be trainable, and they could be taught to carry out much more complex routine tasks. They would also learn quickly to perform their duties with increasing efficiency. Third-generation robots with a brain equivalent to that of a monkey (about 5 million MIPS) could become an important addition to the workplace and the home. They would probably be adopted as part of the family unit and given personal names. Such robots could respond to the way their owners treat them, and many owners would view these machines as 'intelligent'. The Doctor's robot dog K9 was about equivalent to a third generation robot, and by about 2025 such machines could be commonplace.

The fourth-generation robot will have the processing power of the human brain. In some respects it would be considered a 'mechanical human'. And, because the aesthetic design of robots will be developing in parallel with the growth of processing power it is quite possible that within forty years robots will not only have the mental capacities of the human brain but they may also look and feel human.

But would these machines be sentient? Would they be alive?

At the moment nobody really knows. Furthermore, to draw any

conclusions we cannot simply rely on empirical scientific thinking. For most people, satisfactory solutions to this puzzle require contributions from both science and philosophy, a blend of logic and ethical judgement.

Throughout the last few pages I've studiously avoided saying that a machine with a mental capacity equivalent to the human brain would have human intelligence. This is because having the processing power of a human brain is not the same as being human, or even thinking like a human.

There are two great differences between a computer and a human brain. The first of these is that a computer thinks in a linear fashion. That is, it possesses a large number of processors which each work very quickly but do not link laterally. A human brain possesses something like 100 billion processors or neurons, and each of these is linked to hundreds of others. This means they work in a lateral fashion, operating as a network of individual processors.

The result of this difference is that a computer is very good at working out, say, the tenth root of a fifty-digit number, but it would find explaining the plot of *Hamlet* extremely difficult. Very few human beings can solve any mathematical puzzle involving a fifty-digit number, but most people could explicate the plot of *Hamlet*.

The second way in which computers and human brains differ is that a human can gather terabytes of information from their senses (the human interface with the external environment) and store it. This information is then processed in both a lateral and a linear fashion, and we learn to respond to that acquired knowledge. It is a skill called experience. Computers do not acquire experience, as they cannot gather information from the environment.

Robots of course are not the same as computers, and if a sophisticated robot could be designed which is able to access information from its surroundings then it would develop a form of experience. So, in the future, robot intelligence will principally differ from human intelligence in the way each processes information and arrives at conclusions. That said, one avenue of current cybernetics research being investigated by a few adventurous spirits is the idea of the 'neural net'. This involves linking processors in a similar way to the links found between neurones in the human brain. Each processor connects laterally to hundreds of others. This science is in its infancy, but in the future it might lead to the development of robot brains that mimic humans and process information just as we do.

But is it possible that a robot with at least the processing power of a human brain could ever be considered sentient? Could such a robot be considered self-aware, an intelligent living thing?

Well, as we have seen, just trying to define life is surprisingly difficult, and deciding whether or not an artificial intelligence is alive is more problematic still. As for self-awareness, this is such a nebulous thing that scientists, philosophers and theologians have reached no real consensus on the matter even though they have been arguing about it for centuries.

One school of thought is that sentience or self-awareness is a by-product of a certain level of brain capacity, a gestalt perhaps, a spin-off of intelligence. If that is the case then a robot with a sufficiently large brain should acquire self-awareness. But, it could be argued, this gestalt is only possible with certain types of 'brain'. Perhaps only human-type brains, the lateral processing machine we have inside our heads, generate the form of intelligence that leads to self-awareness. But then even this idea is not entirely watertight. Would a robot with a brain wired as a neural network that simulates the human brain not be sentient? Could robots ever be able to acquire intuition? Could they ever have gut feelings? Could a robot ever act upon a hunch? And from these considerations we are inevitably led to the questions: could a robot feel emotion? Could a robot acquire a sense of self-preservation? Could a robot love?

These are all extremely difficult questions to deal with, and I suspect that we will only ever begin to formulate answers when we have some experience of the matter. Perhaps in half a century from now we will be forced to address these questions.

Today, robots are approximately at the same stage as the aeroplane at the start of the twentieth century. Robots are little more than toys, but they have the potential to change our society radically. Currently, more than a million robots are in use globally. There are 3,000 robots working under the sea, nearly 2,500 demolition robots and some 1,600 surgical robots employed in hospitals around the world. The number of robots in our homes has increased from 12,500 in 2000 to almost 500,000 today. Relatively few of these are to be found in the UK or the US, and more than half of the world's working robots have found gainful employment in Japan, where the cybernetics industry is booming and robotics research leads the world.

It is likely that by the middle of the 21st century many hundreds of millions of robots will be in use in a vast range of applications. As technology improves, these robots will not only acquire at least the level of processing power we possess in our brains, but they will start to look more and more like us. It is inevitable that, as the computer experts push forward the power of artificial brains, designers will find ways to simulate skin, to reduce the noise of artificial joints, to manufacture small power supplies. By 2050 a robot might look, feel and communicate so much like a human that they could almost be passed off as one.

When this is possible our relationship with robots will begin to change radically. Robots will become members of the family. Children who have never known a world without robots will communicate with their domestic machines as if they were sentient. Humans will become attached emotionally to their mechanoids and many will begin to propose that robots should have rights, share civil liberties and be responsible under the law. It is quite likely that the evolution of robots into androids that speak, smile, share concerns and express emotion (albeit pre-programmed and simulated emotion) will be a vehicle for social change comparable to the printing press, the car, the television or the internet.

Fifty years from now we may well be presented with a set of social dilemmas few of us today would consider to be realistic. For instance: how human is our home robot? Does shutting a robot down constitute assault? Would a married person having sex with a robot be committing adultery? These may seem like ephemeral issues now, but children who are starting school today may well have to deal with such matters when they are parents and raising families. Indeed, these dilemmas could become as important to those living in 2050 as arguments over stem-cell research, abortion or cloning are to us.

What this illustrates is that the answers to the thorny questions I postulated earlier may only ever be dealt with when they arise. To answer questions about the sentience of artificial life or whether a robot will be capable of self-awareness is almost impossible for us to consider now, but the need to find answers may well become unavoidable.

And if all this seems to be an unnecessary complication for what is ostensibly just the development of a machine, consider the following scenario. Imagine you have one of the 100 billion neurones in your brain replaced with a microchip. Would you feel any different? Almost certainly not. What if 10 brain cells were replaced? 100? A million? All of them? As they were replaced would you gradually lose your identity? Would your self-awareness go and be replaced by some other form of consciousness? Would you be dying or simply changing? Finally, would you still be a human being with rights and responsibilities, or a machine with none?

One last question to consider, which may well become the most significant issue for future generations is this: by creating an android with a brain capacity at least as large as ours, a being that will probably be stronger, faster and longer lasting than the average human being, are we not creating our own nemesis?

Today, this is another imponderable. All we can propose now is that the development of robotics has to be monitored carefully and regulations and guidelines are required to ensure that research is responsible. But ignoring the possible dystopias and putting an optimistic spin on the matter, in the future robots might just offer human beings a better world in which all the dirty jobs are taken on by non-sentient automata, leaving us to apply our lateral brains to higher things. Naïve perhaps, but why else pursue robotics at all?

The answer to this leads to one further consideration. As robots are shaped in our image and made to look more and more like human beings, we will also start to become more like robots. For, as I will explain in the next chapter, the age of the cyborg is almost upon us.

9

To Live For Ever

How Close Are We to Regeneration?

I plan on living forever. So far, so good.   Anonymous

For many people one of the most enduring fantasies is to live for ever. Second to this is the dream that we could extend our lives into healthy and happy old age; looking 25 with a birth certificate that marks us down as 200. But how close are we to the reality of extending our lifespan to centuries or, indeed, living for ever?

One of the best-known characteristics of the Doctor and indeed all Time Lords is their ability to regenerate. The Doctor has undergone eight regenerations (and is about to go through a ninth to take on his tenth incarnation, played by David Tennant). For Time Lords, regeneration is the way in which they can sustain their lives almost indefinitely. Sometimes the regeneration is spontaneous, on other occasions it is planned.

The closest Nature comes to any form of regeneration is the metamorphosis of the caterpillar into the butterfly. But in at least two ways this is different to the form of regeneration experienced by the Doctor. First, it is much slower, but more importantly, perhaps, the transformation of a caterpillar to a butterfly or a moth is simply a step in the lifecycle of the creature and not a way to replicate the same form of being.

Nature has provided humans with a lifespan much shorter than that of the Time Lords or indeed of most of the aliens encountered in *Doctor Who*, but through the simple refinements of better hygiene, improved medical facilities and improved diet, today we live considerably longer than our ancestors did. But we all cherish life, and it is only natural that we should want to apply our improving knowledge of science and the technologies that spring from it to both lengthening and improving our lives.

No humans regenerate naturally, and one might imagine that the Time Lords do not regenerate naturally either but have somehow changed themselves, by the application of advanced technology, into beings that can do so. Such intervention is the only way in which human beings will ever live for centuries or millennia, and today scientists can visualize two different ways in which this may be possible. One involves biochemical intervention, while the other requires the transformation of humans into cyborgs.

Current theories suggest that the mechanism behind ageing involves a complex chemical process using oxygen. Oxidation in the cells of our bodies produces a chemical entity called a 'free radical', which can severely damage cells and disrupt metabolic processes. By using anti-oxidants, scientists working under laboratory conditions have shown that the lifespan of simple organisms (including fruit flies) can be lengthened considerably. Of course, there is a big difference between the genetic complexity of a fruit fly and a human being, but this research at least offers promise for the future.

More exciting is the possibility of using genetic engineering to reverse ageing. In the 1960s it was discovered that the cells of newborn babies were able to divide 70–80 times, whereas cells taken from a mature adult only managed 30–40 divisions before dying. This implied that all cells have a built-in life expectancy, but the source of this phenomenon was a mystery.

A decade later researchers had found the cause. The ends of each chromosome in our cells are capped with what is called a 'telomere', a tip made from a collection of proteins. During the course of an organism's life the telomeres are worn down, and after a while the chromosomes begin to stick together, which eventually causes the cell to die. A further ten years of research led geneticists to the discovery of telomerase, an enzyme which reverses the breakdown of the telomere. A further decade on, in the mid 1990s, scientists working at McMaster University in Canada made the startling discovery that many cancer cells contain telomerase, which allows them to survive much longer than host cells.

During the past decade a huge amount of research has gone into investigating the role of telomeres and how manipulation of our genes could help to lengthen human life. In the lab, cultures of human skin cells have been made to live almost twice as long as expected, increasing cell division from 50 to 90 times (more than the number expected from the cells of a newborn).

This research may eventually lead to ways in which the process of ageing could be reversed, but this still offers only a small part of the picture. It is all very well finding ways in which human beings might be able to increase their life expectancy, but this does not necessarily guarantee healthy long life.

One way healthy old age could be achieved is by using cloning technology. At the moment, progress in this science is being held back because of moral objections, but it is the most promising way in which biology can radically alter both human longevity and physical wellbeing during an extended lifespan.

Cloning has been a science-fiction favourite for many years, but it has only recently begun to emerge as a respectable and potentially epoch-changing science. As a theoretical possibility, cloning has been taken seriously since the earliest days of genetic research, and as long ago as the 1970s it entered the public arena. It has been the subject of several novels (probably the most famous of which is *The Boys from Brazil*, in which the Nazi Josef Mengele clones Hitler's genes in an attempt to produce a master race), and more recently it has been a theme in movies such as *Gattaca, Stepford Wives* and *Multiplicity*.

However, only a few decades ago most geneticists considered practical cloning to be a long way off. In part because of this, for many years cloning remained on the scientific sidelines. But in July 1996 the world was shaken by the news that a little-known British embryologist, Ian Wilmut, had become the first person to successfully clone a mammal from adult cells. A sheep known as Dolly had been 'created' in the lab.

News of this breakthrough was immediately splashed across the newspapers of the world, and it soon became the subject of books and television documentaries. But what exactly is cloning, and how could such work as Wilmut's offer the potential to live 200 years in rude health?

Cloning is a process via which the genetic material of an egg is completely removed and replaced with the entire genome (a complete set of genes) from a donor, the individual to be cloned.

Simple types of cloning have been possible for some time; indeed, before producing Dolly, Wilmut and his team had successfully cloned two sheep, named Megan and Morag. But these had been produced by splitting a single embryo, creating two identical copies of the same sheep. The method used by Wilmut was different from earlier experiments in two distinct ways. First, until Dolly, only relatively simple organisms had been cloned by implanting genetic material into a host egg (rather than splitting an already fertilized egg). The list of cloned creatures included bacteria, plants and a few frogs. The second and even more important factor distinguishing Wilmut's achievement from all other previous experiments is that he used genetic material taken from an *adult* animal, a mature sheep, and implanted this into a recipient egg.

This achievement changed genetics radically. For the very first time, a scientist had brought to reality the 'science fiction' idea of cloning – taking genetic material from an adult animal, removing the material already present in a recipient egg and replacing it with 'foreign material'. What this means, if we extrapolate the science only a tiny degree, is that an adult human could have genetic material removed. This could be placed in a host egg and a new human could be grown in the lab. This human would not be merely a twin or close relative of the original human, it would be an identical, *exact* copy of the original, possessing the same genome.

The process enabling this to happen may be broken down into seven steps:

1. Remove some donor cells. Place these in a glass dish and label it *clone*.
2. This is what may be considered the most crucial step: put the cells into a state of 'hibernation'. To do this they are starved of nutrients, which means they stop multiplying and enter a state in which

they can 're-programme'; they are 'tricked' into thinking they have returned to an embryonic state and become ready to multiply once the new DNA has been added to them and the old DNA removed. No one really knows how the starving of cells sends them into this state.

3. Give a special cocktail of drugs to a large group of women to push them into a state of 'super-ovulation' in which they produce large numbers of eggs in a short space of time.

4. Harvest the eggs and remove their nuclei. The nucleus is the nerve centre or 'brain' of the cell and contains the twenty-three pairs of chromosomes that constitute the genome of the human. This stage of the process requires great dexterity, because we must not lose or damage the rest of the cell's material within the outer cell membrane – the cytoplasm.

5. Place a nucleus from the removed cells of the donor to be cloned into the recipient cell in place of the removed nucleus. Fire a pulse of electricity into the cell. This facilitates the acceptance of the new material in the recipient cell. No one knows why a pulse of electricity will do this, but it does.

6. Place this fused egg into the womb of another woman who acts as the surrogate mother for the clone.

7. Wait nine months. The expected success rate for mammals has been placed by Wilmut's team at about 1 in 300. This figure is based upon his own experiments using sheep. No one knows what the success rate would be in human cloning.

As soon as news of Wilmut's cloning of Dolly was announced, fears were voiced about the possible application of this work to the cloning of humans, and the United Nations has placed a 'non-binding' ban on any form of research involving human cells. This places limits on many scientists, but some countries have chosen to ignore the ban. Britain is one of the few places where research into cloning is supported by government. Only research into human 'somatic cells' (cells involved in reproduction) is prohibited.

Ethical concerns have held back research in the United States, where any form of cloning experimentation involving human cells is banned under legislation introduced by the Bush administration. The natural consequence of this is that British scientists have become world leaders in cloning research, and even if the Democrats (who are more likely to favour cloning) are elected in 2008, American research in this incredibly fast-moving field might have fallen so far behind that it would be extremely difficult for them to catch up.

One aspect of cloning is the idea that it might be possible to duplicate a living or dead human. Science-fiction stories abound of wealthy egotists who want to 'live for ever' by reproducing their bodies as more-youthful doppelgängers into which they can somehow implant their personalities and memories. Other stories (such as *The Boys from Brazil*) offer the many potential dystopias that could conceivably arise from the misuse of cloning: these include clone armies, Frankenstein scenarios and many other imaginative possibilities. But making new humans that have not been born the conventional way is only one possible application of this technology. In many ways, a far more exciting and potentially much more beneficial use for cloning is the ability to create cloned organs and other body parts.

A refinement of the seven-step procedure described above is a more versatile study called stem cell research. Stem cells hold vast promise for treating an array of diseases from diabetes to Parkinson's disease, and they could also offer the potential for increasing human lifespan considerably. The remarkable thing about stem cells is that they possess the ability to grow into any type of human tissue, and scientists hope to be able to direct the blank cells to grow into specific cell types needed for transplant.

Stem cells can be found in adults, but it is believed they may not be as versatile as those found in embryos. Rather than using adult stem cells it is far easier to create an embryo from a patient so that extracted stem cells may offer a perfect transplant match. What this means for the future is that we could each have spare parts ready for transplantation even before they are really needed. We could have a body-part replacement session every 30–40 years, say, to keep us in good shape.

Those opposed to stem cell research claim it is wrong to use embryos (which are kept frozen at just a few days old and consist of nothing more than a tiny collection of cells) for experimentation. They base their feelings on the idea that life is created at the moment of conception and that even a collection of a few dozen cells is 'a living human'. Although there is absolutely no evidence to support this idea, those who claim to be 'pro-life' have a great deal of influence, in part because they have inevitably become aligned with the Christian Right (especially in George Bush's America) and they have a loud voice. It is the influence of the anti-cloning groups that has led to the current ban on cloning in the United States.*

Meanwhile, in May 2005 British scientists were the first to announce they had successfully cloned a human embryo that could be used as a source of stem cells. In the same week, South Korean researchers became the first to produce tailor-made stem cells that could be used to produce replacement organs for humans.

It is inevitable that most societies will adopt cloning as a technique for creating organs and other materials that may then be used to cure disease and to lengthen the lives of human beings. Never in human history has a new technology been stopped completely for ethical reasons, and once the genie is out of the bottle there is no way to get it back in. Cloning may be used to replicate humans, but I'm inclined to take the view that, after a few illegal experiments that succeed in producing human clones, it will be seen as a rather pointless exercise. The alternative scenario of cloning body parts is far more attractive, because it offers most of the things human clones could offer without the social, ethical and practical problems.

I began this chapter by proposing two routes to longer life and the possibility of people living to be fit and healthy bicenturians. But for many researchers the notion of humans living to the age of 200 is actually rather conservative. Some visionaries foresee a future in which we have the potential to become almost immortal, and they claim that living for thousands of years is not only theoretically possible but will be practically achievable in a century or two. This

---

* The law against cloning in the United States does not extend to private research institutions, and some innovative work is being done in California, where the state government is encouraging stem cell research in the private sector. In May 2005 a bill was put through Congress to allow stem cells to be extracted from the umbilical cord so as to bypass the need to destroy embryos. However, many scientists believe this will not provide the flexibility of the standard procedures.

could be done by replacing our flesh-and-blood bodies with artificial counterparts; in other words changing ourselves into cyborgs. This is an idea that is certainly not new to science fiction. From Fritz Lang's *Metropolis* to *Star Trek: Voyager*, cyborgs have been written about and portrayed on film in many different ways. We've seen the warrior cyborg (*Terminator*), the sexual-fetish cyborg (*Star Trek: First Contact* and the character 7 of 9 in *Voyager*). There is the femme fatale cyborg (Pris in *Blade Runner*), the hero cyborg (*The Six Million Dollar Man*), the everyman cyborg of William Gibson's cyberpunk worlds, and more prosaic perhaps is the body art 'cyborg' of the present-day piercing and tattoo fetishist.

However, the concept of the cyborg was first introduced as a scientific possibility (rather than merely a thing of fiction) by the cybernetics researchers Manfred E. Clynes and Nathan S. Kline in a paper entitled 'Cyborgs and Space', which they wrote for the journal *Astronautics* in 1960. This was the first time the term 'cyborg' was used as a shortening of 'cybernetic organism'. Since then, thousands of papers have been written on the subject, and the idea of blending humans with machines has become an accepted discipline that many believe will soon begin to make a profound impact upon our lives.

The simplest forms of cyborg technology are the prosthesis and the skin graft. Wooden legs were used in ancient times, but during the last fifty years artificial limbs have become more and more sophisticated, to the point where, today, patients can be fitted with electrically powered hands that allow them to open doors, operate a computer and turn the pages of a book. The electronic circuitry that makes this possible has been miniaturized, and processors operate the cables needed to simulate normal movement while tiny batteries placed in the thumb power the prosthetic hand.

What makes these devices really special is that the signal to move the hand comes from nothing so crude as a switch. Instead, an electrical trigger from the brain of the patient activates a microprocessor which then operates the fingers and the thumb of the hand. The prosthesis literally replaces the lost hand, because when the patient wants to move a finger or a thumb or to grip an object they simply think it and the hand moves.

Colonel Steve Austin, a.k.a. The Six Million Dollar Man, had a great deal more than a hand replaced, and his prostheses were built to provide him with superhuman strength and senses. This slice of science fiction was way ahead of its time and in some respects it was quite prophetic.

The Six Million Dollar Man was first screened in 1974, but even now, more than three decades on, the surgical techniques and micro-engineering skills required to carry out the extreme makeover Colonel Austin experienced remain far off. There is also the fact that when it finally does become possible to carry out such surgery, the bill will come to a lot more than six million dollars!

Today, a good portion of the research being conducted to create 'super prostheses' and other related medical wonders is funded by the military establishment. For some time fighter pilots have used specially designed goggles that display a vast array of detailed data projected on to tiny 'screens' directly in front of the pilot's eyes. The next step will be to send this data directly to the pilot's optic nerves. Later still, technology could be developed to simply send the infor-mation directly to the relevant part of the flyer's brain.

A direct spin-off from this is the application of such devices to eye surgery. In 2000 a blind patient was given partial sight by hooking up his visual cortex to a minicomputer attached to a belt worn around the waist. Using this system, a tiny video camera sends signals to the minicomputer. Special software cleans up the image and simplifies it before passing on the refined signal to a set of sixty-eight platinum electrodes implanted into the visual cortex of the patient. Using this equipment the patient (who had been totally blind for more than twenty-five years) could read two-inch tall letters from a distance of five feet and was able to see well enough to navigate the New York subway unassisted.

A further refinement of the pilot's goggles is a new flying suit studded with sensors wired up to a powerful but very small computer. This allows the most inexperienced pilot to fly a plane by passing on data from their body movements to a computer which translates the information into instructions to fly the plane properly. This is an example of a wearable computer system, where the wearer melds with the suit and enhances their abilities and sensitivities. Very soon, wearable computers will be commonplace and link up with the fashion industry to create machines that provide entertainment (music, video, virtual experience, sexual pleasure), information (linked up permanently to the internet and videophone connections) and enhanced body function (to help overcome fatigue, provide protection from extremes of temperature, and to even allow greater strength and speed).

This technology will be welcomed by anyone with a physical disability. The age when people with damaged legs need to sit in a clumsy wheelchair will be forgotten, and those who have lost the use of their arms will be able to function normally. Today the computer power needed to simulate sight is almost achievable, and as new ways are found to link computers and brains, many blind people could have their sight restored by surgery. In a similar way, those with impaired hearing or speech impediments will soon be able to use machines to enable them to hear and speak.

Some see this as just a beginning. Those with physical disabilities may be the first to benefit from augmentation, but many researchers in this field envisage a future in which we may be capable of changing our bodies almost beyond recognition, lengthening our lifespan almost indefinitely.

Cyber-evangelists, sometimes referred to as 'post-humanists', are almost ridiculously optimistic when it comes to their ideas for the future of humankind. They talk of ways in which nanotechnology would transform our natural bodies into something almost unrecognizable, turning humans into superhumans. In other words, it is seen as quite possible, theoretically, to transform men into Cybermen.

Even enthusiasts of this science admit this isn't going to be easy. We are talking about a complex and radical process, the metamorphosis of our bodies from the flesh-and-blood models we are born with into cyborgs. This is a process dubbed 'accelerated and directed evolution', and it will take time. Even so, it is certainly not impossible. Indeed, proponents of accelerated and directed evolution see almost no limit to what is possible. But how realistic is this? How far could this process be taken?

Today it is possible to replace hearts and lungs, kidneys and livers using donor organs. It is also possible to replace the heart with a mechanical device, and the functions of the kidney can be carried out by a machine during dialysis. Surgeons are able to save limbs by reattachment, and prosthesis technology is improving rapidly. But imagine doing away with the heart altogether. Imagine replacing our skin with a better material, or bypassing the need for food.

All of these changes are possible using nanotechnology. According to some researchers, it is possible to replace blood cells with nanites that carry out the same functions and more. They envisage nano-machines carrying oxygen and nutrients to the cells of the body using the bloodstream as a highway. Each of these tiny machines will be self-propelled so there would be no need for a heart. After all, the cyber-evangelists say, the heart is a pretty vulnerable piece of machinery which often breaks down long before other systems or organs do.*

If this sounds ridiculously far-fetched, we should get this notion into perspective. First, a century ago the idea that a human heart could be replaced with a mechanical one would have been laughed at. Furthermore, designs for nanobots that travel the blood stream are already on the drawing-board. 'Blood-stream-based biological microelectromechanical systems' (or bioMEMS) will be in use within a decade to regulate the biochemicals in the blood and to act as a diagnostic tool.

By using nanobots to supply nutrients we will no longer need to eat or to excrete. Before this stage is reached we will be able to control the processes via which we take up nutrients in our bodies and distribute the energy and the waste. This will mean that we could eat what we like and never get fat. Later, we could eat merely for pleasure, just as sex is primarily recreational.

Taking things a step further, the skeleton could be replaced with an infrastructure made from longer-lasting and hardier materials. This will mean an end to osteoporosis and arthritis and fractures will be far less common. It will also mean that we could redesign the body completely.

---

* The Time Lords have clearly not adopted this particular aspect of augmentation, because all members of their species have two hearts.

Some science-fiction writers and futurologists have pondered the possibility of transforming human beings into entirely different animals. For example, could redesigned humans fly? Could a human being be changed into an angel? A humanoid with massive wings, perhaps?

This is fun to imagine, but how difficult would it be? Well, the first thing to realize is that sticking a pair of wings on a human frame just isn't going to work. We only need to consider old, flickering movie footage of the Wright Brothers' predecessors attempting heavier-than-air flight to see the truth of this. Those crazy experimenters nose-dived into the sea or collapsed under the weight of their machines because no normal human physique could act as the support structure for suitable wings. As long ago as 1680, a medic and flight enthusiast named Giovanni Borelli showed that human muscles were far too weak to move the large surface area needed for effective wings, and that in order to provide the necessary power the human heart would need to pound at a rate of 800 beats per minute.

However, by transforming the skeleton and using nanobots to replace the heart, it would be feasible to turn a human being into an angel. Using advanced materials such as a composite made from a network of carbon atoms called 'buckminsterfullerene', the bones of the body could be made much lighter but thousands of times stronger. The nanobots could supply oxygen to the cells at a greatly enhanced rate, and the muscle structures needed to flap the wings could be designed to accommodate.

Others undergoing augmentation surgery may wish for fish scales or feathers, snake skin or bear-like fur. This would be possible by replacing the skin with new materials which simulate the epidermal layers of real (or indeed imaginary) animals; the choices would be endless.

So much then for physical transformations. But what about accelerated evolution for the brain? How far could we take this line of development?

In the near future it will be possible to create links between what is called 'wetware' (biological neural structures) and hardware (electronics of one form or another). This is precisely what is happening in the latest prostheses, which allow the patient to move their artificial limbs by just thinking about it. It requires only a small step to allow humans to interface more fully with machines and operate them with their minds. This is a process dubbed 'teleoperating', and it takes little imagination to see how useful this would be to those with physical disabilities.

Taking this a stage further leads us to the possibility of linking our minds with the full functionality of a computer, effectively expanding our minds to incorporate the electronic device. This is still a long way off, but it is an area of research which is advancing rapidly, with plenty of military and industry funding.

In 2001, scientists at the Max Planck Institute for Biochemistry in Munich became the first to blend wetware and hardware when two snail neurons were linked to a microprocessor. Researchers found they could pass electrical signals between the neurons and the chip and also between the two snail neurons. The next step is to use many more neurons to form a network and to see if information from a processor could be stored in the wetware network. With this technology the human mind could be expanded to an incredible degree. Imagine linking your mind directly with the internet. In effect you could have access to all the information ever known to humanity.

Another aspect of this is the way such technology could be used in combination with physical augmentation to allow humans to live practically indefinitely. The post-humanist scenario of replacing body parts to construct superhuman bodies, or the more prosaic techniques of organ replacement using cloning, would be combined with a continual upgrading of our neural systems.

In theory, at least, it will be possible to implant a processor into the brain of a newborn (or even *in vitro*). This could act as a backup drive in which every experience and emotion is stored, every byte of information picked up by the senses recorded digitally as well as biologically. If, later, as an adult the subject falls mortally ill or is involved in a deadly accident, the chip could be removed and implanted in a new body, effectively providing immortality.

Another spin on this is the idea that in the future we will be able to experience being someone else. We could swap processors to have access to each other's minds or simply draw experiences from a vast databank of backup drives for any human who wanted to expose themselves in this way. Yet another seemingly fantastic possibility is the idea that all human experience could be melded into a single entity in cyberspace – in effect the creation of a superrace, like the Borg perhaps, a hive or a colony.*

These are all fascinating ideas, and there is nothing scientifically unfeasible about any of them. Some of these visions are thousands of years away from ever being possible, but they are certainly not impossible. The question is though, would most of these imaginative scenarios be desirable? Should we create them simply because we can?

It seems to me that there will come a point where the advantages from this technology are outweighed by the disadvantages. Most people would like to lead long, happy and healthy lives, and the prospect of 200 years of physical and mental fitness would appeal to many people. But would you really want to live for ever? Would you like to have your body completely replaced by machine parts, however advanced they might be?

Post-humanists see death as avoidable, a thing that may be kept at bay indefinitely, and perhaps they are right. It has been pointed out, quite rightly, that death is a terrible waste. According to one estimate, approximately 55 million people die each year (a little less than the population of Britain). Only 3 million of these are caused by human action, mostly wars and other violence, accidents and suicide. The remaining 52 million deaths are by 'natural causes', principally illness of one form or another.

In the eyes of post-humanists, this loss is an outrage. The great pain that death causes loved ones left behind is one thing, but death is also a waste of the years a person spends acquiring knowledge and experience. It has been estimated that each death represents a loss of information equivalent to a book. This means that each year the information loss to humankind is comparable to that contained in the British Library.

But suppose the dreams of the cyber-enthusiasts could be made

---

* The Time Lords are said to have created something similar called The Matrix, which is a 'superbrain' containing the sum total of Gallifreyan knowledge.

real, and human beings could live far longer. What would such longevity do for the future of humankind?

The obvious consequences would be either extreme overpopulation or the creation of a world in which few people bred and the average age increased significantly. Neither prospect is very appealing. Overpopulation is a nightmare scenario that has been an influential sociopolitical factor since the last quarter of the 20th century, and it is a matter of growing concern. The flipside of this, a situation in which the average age of the population rises considerably, leads to the possibility of stagnation, creating a society which has lost its impetus.

One possible get-out clause is the idea that, as we develop new technologies that enable us to live longer, other scientific advances will provide the opportunity for interplanetary travel and, eventually, interstellar colonization. This would give the human race space to expand.

By pursuing this line of reasoning, we return to the start of this chapter and a consideration of such beings as Time Lords, Daleks and Cybermen. The Time Lords are members of a civilization that is hundreds of millennia more advanced than humankind. They have mastered some technique to give them greatly extended life through regeneration. Perhaps their culture was saved from stagnation by the discovery of time travel and interstellar travel which allowed them to explore the universe.*

The Daleks are beings who originated on Skaro, and they rely entirely on machines designed for them by the Kaled scientist Davros. But for present-day human cybernetics experts, the choice of mechanisms employed by Davros is puzzling. Why, for example, would such an advanced race implant themselves in such ungainly and impractical shells?

The Cybermen, beings who originated on the planet Mondas, are far more like it. Humanoids possessing superhuman strength and seemingly powerful minds, the Cybermen are, however, beings who have lost all trace of emotion or any semblance of their biological

---

* It is interesting to note, though, that there does appear to be something fundamentally wrong with their civilization, a problem that may be linked to their virtual immortality: 'Time Ladies' are never mentioned in *Doctor Who*.

origins. Cold and clinical, they are intent only upon destruction and domination.

Of course, imaginary scenarios filled with such creatures as Cyber-men or Daleks represent ridiculous extremes, and they are neither necessary nor at all likely. But this is the way of science fiction. Writers are often compelled to take a contemporary technological strand and extrapolate into a future that is either black or white. This generates old-fashioned science fiction in which the Earth is invaded by flesh-eating aliens, the planet is stripped of resources, or we face a future in which robots destroy mankind or instead we evolve into soulless cyborgs.

Rather than being black or white, reality is much more likely to be grey, or, I like to think, technicoloured. Humans muddle through, make mistakes and find ways to solve problems. We do our best and somehow find a path for civilization to follow, almost always without recourse to ridiculous extremes.

Epilogue

Gallifreyan Magic

*Why shouldn't truth be stranger than fiction? Fiction, after
all, has to make sense.*   Mark Twain

The Doctor is a weird blend, a non-human, time-travelling being who
has amazing technologies at his fingertips, but who insists upon using
something so prosaic as a sonic screwdriver. The title for this book
comes from a quip of the Doctor's which sums up his eccentricities
perfectly. In response to being asked how he came to defeat one of
his opponents, he replied: 'Well, to be fair, I did have a couple of
gadgets he probably didn't, like a teaspoon and an open mind.'*

Like a lot of science fiction, *Doctor Who* is a blend of the possible
and the impossible; but it would be more accurate to say that it is a
confection of science, science fiction, fantasy, the occult and, yes, the
mundane.

THE TARDIS

The Tardis is the most innovative and imaginative thing about *Doctor
Who*. The Doctor's machine was an old model, a Type 40 to be
precise, and its shape, that of a 1950s London police box, is wonder-
fully British and idiosyncratic, but it fits seamlessly with the rest of
the *Doctor Who* mythology.

The reason for the outward appearance of the Tardis is that its

---

* This was in the 1979 episode *The Creature From the Pit* starring Tom Baker.

chameleon circuit was irreparably damaged on Earth in 1963 when, on a trip to early sixties London, the Doctor instructed it to take on the shape of a contemporary police phone box. The chameleon circuit is a piece of equipment built into every Tardis which gives it the ability to take on any shape the owner desires, allowing it to merge into the background. This is obviously very useful as the Doctor (or any Time Lord) can leave their machine where they like and it will go largely unnoticed.

More important than this, though, is the very idea of making the Tardis transdimensional. The fact that it is larger inside than outside is pure psychedelia, years ahead of its time. Considering this was a concept dreamed up in the early sixties, it was pretty left-field even for science fiction.

So, is such a thing as transdimensionality actually possible?

The answer to this is almost certainly 'no', so that, sadly, the thing about the Tardis that makes it so special is also the thing that is most far-fetched about it. From what we understand of physical laws, it would appear that transdimensional abilities (as demonstrated by phone boxes that are no more than a few feet square on the outside but cavernous inside) are actually impossible.

And what of the chameleon circuit?

Well, although modern science has no exact analogue of such an object – scientists are not able to completely transform the shape of things while retaining their functionality – it is though at least possible to give the illusion that a shape has changed.

The idea of the shape-shifter is one that is popular in occult circles, and a great deal has been written about beings that can apparently transform themselves from one shape into another. It is also an idea that has been adopted by science fiction and used frequently in *The X-Files* (remember Eugene Tooms from series 1?) as well as in movies such as *Terminator III*.

In Nature there are creatures that are capable of changing their shape and colouring to a very limited degree. Snakes are incredibly flexible and versatile in the shapes they are able to adopt, and mice and cats each have skeletons which allow them to squeeze into very small spaces with relative ease. Colour change is more common in Nature, and this ability has many uses. Some animals, such as the chameleon itself, are able to make themselves invisible to predators by merging into the background. Other animals use colour to make themselves appear bigger and more frightening to a rival or predator (certain species of sea slugs and frogs do this, of course). A third group uses colour to impress the opposite sex (the most obviously example being the peacock).

In 2004, scientists working at the National Institute of Advanced Industrial Science and Technology in Tsukuba, central Japan, announced the prototype of a shape-shifting robot they called ATRON. This has been constructed from over 100 separate modules which can rotate and realign themselves with each other. This means ATRON can slither like a snake, walk like a biped or quadruped, or can even roll around in the shape of a sphere.

A variation on this is a robot being developed by NASA called TETwalker (an abbreviated form of tetrahedral walker). In one plan for a future mission to Mars many of these tiny robots would be sent across interplanetary space in a single probe. The TETwalkers would then be released into the atmosphere of Mars and parachute to the surface where they could perform a range of tasks by taking on different shapes, starting with the tetrahedron or triangular-based pyramid shape. Like ATRON, such robots could operate in many different terrains: they could roam deserts and mud flats, climb mountains, float on water and squeeze into tiny fissures.

An extension of the idea of shape-shifting machines is stealth technology. This works on the principle that an object may be made almost invisible to radar and other forms of electromagnetic radiation. To achieve this, an outer shell of reflective plates is used and the object to be disguised is coated with materials that absorb radar signals. Most conventional aircraft have smooth edges to make them aerodynamic, but this shape also acts as a very efficient radar reflector. The round shape means that, no matter where the radar signal hits the plane, some of the signal gets reflected back. A stealth aircraft or ship has completely flat surfaces and very sharp edges, which cause the signal to reflect away at an angle.

The overall result is that a stealth aircraft like the F-117A built by Northrop can have the radar signature of a small bird rather than an aeroplane. At sea, stealth ships such as the Kongo Type destroyer currently in use with the Japanese navy or the Royal Navy's Type 45 destroyers and Type 23 frigates are revolutionizing the nature of marine warfare. Designers have reduced the radar cross-section of these ships by up to 99 per cent. This does not mean they are 99 per cent invisible to enemy radar, but they are very much harder to detect than conventional ships. Using the most sophisticated radar equipment available, these stealth vessels can detect an enemy vessel from a distance of 100 km, but they remain undetectable to within 30 km of a target.

CRYSTALS

Crystals feature quite extensively in *Doctor Who*. There are crystalline life forms such as the Krotons, the Kastrians and the Krargs. There have also been stories in which the Doctor has met aliens who use crystals of power. One example was the White Guardian, who offered enlightenment in the form of a crystalline cube. A race called the Borad used crystals to facilitate time travel, and the third Doctor, played by John Pertwee, encountered the high priests of Atlantis, who had captured the essence of the powerful Kronos and kept it in a crystal.

Most importantly, in the back story of *Doctor Who* we learn that some unspecified form of crystal acts as a power source for the Tardis, and it is implied that the Time Lords long ago discovered useful properties of certain exotic forms of crystal. Indeed, Gallifreyan legend relates how the scientist Omega (the Time Lord who first achieved the ability to travel in time) used a crystal to power his time capsule.

Other science-fiction stories also feature crystals. The Starship *Enterprise* is powered by dilithium crystals, and crystals are often used as vehicles for concentrating energy beams or for storing strange energy fields. Like the idea of shape-shifting, the function of the crystal is also a staple of the occult tradition, an interest that probably has its roots in ancient times when the first alchemists (whose stock-in-trade was the creation of crystals) tried to produce the mythical Philosophers' Stone.

Crystals also intrigue the modern scientist. The study of crystalline structure is a large and multifaceted one that overlaps with many areas of research, some of which have huge significance to technology. Many types of laser operate by use of a crystal that is employed to split electromagnetic radiation into suitable components and focus the beam. Semi-conductors incorporated into most electrical devices rely on the properties of crystals of silicon and germanium, and still other types of crystal are used in optical devices; in thin-layer technologies which allow coatings just a few molecules thick to be applied to surfaces; and in the heart of liquid-crystal displays.

So far, though, crystals have not been put to use as a way of creating or focusing large amounts of energy in the way the Gallifreyans are capable of doing. The most powerful energy source in use on Earth today is nuclear fission, which provides around 15 per cent of global energy needs. But, as I mentioned in Chapter 4, in the near future it might be possible to generate a substantial proportion of global energy requirements using some form of nuclear-fusion technology, and according to some recent research, this method may be brought closer to reality using crystals.

According to a paper published in *Nature* in May 2005, scientists at UCLA and in Glasgow have succeeded in producing a refined form of nuclear fusion on a lab bench. Brian Naranjo and Seth Putterman at UCLA and Jim Gimzewski from Glasgow achieved this using deuterium (a heavy isotope of hydrogen, which is also used in conventional fusion processes). The method relies on using the strong electric field generated in what are called 'pyroelectric crystals'. These are materials that produce electric fields when they are heated.

In this new study, the researchers concentrated this electric field at the tip of a tungsten needle connected to the crystal. In an atmosphere of deuterium gas, this produced positively charged deuterium ions, which were then accelerated using a high-energy beam.

Scientists can tell when fusion has occurred because the process produces large numbers of neutrons as a by-product. When the beam produced from the pyroelectric crystals in this new experiment made contact with a target composed of a substance called erbium deuteride, Brian Naranjo and his colleagues found that neutrons were being emitted from the target and that these possessed precisely the energy they would be expected to contain if they had been generated by the nuclear fusion of two deuterium nuclei.

When interviewed about this discovery, Professor Putterman claimed that he had originally wanted to call the technique 'crystallic fusion', but was then told by his children that Buzz Lightyear uses this very term in *Toy Story*, when he declares: 'Are you guys still using fossil fuel? Haven't you discovered crystallic fusion power yet?'

The property of pyroelectric crystals used in this experiment was known about by the Ancient Greeks 3,000 years ago. In some respects it can be considered an electrical analogue of a permanent magnet and if you heat or cool these crystals you can build up a very large charge and a substantial electric field. However, the most striking feature of Naranjo, Gimzewski and Putterman's discovery is that large amounts of neutrons are being produced at low temperatures (as low as room temperature), which means that fusion is occurring with amazing ease.

This could be a very exciting development, but, being aware of the shaky history of cold fusion (as described briefly in Chapter 4), scientists (not least those involved in the discovery) are treating their breakthrough with caution.

RESONANCE

One of the most important properties of crystals is their ability to resonate. Crystals are solid structures with very regular molecular frameworks. A simple example is a salt crystal, made from just two types of ion, the positively charged sodium ion and the negatively charged chloride ion, which form a cubic arrangement. Many crystals have far more complex structures, but most sorts of framework can be made to vibrate at specific frequencies.

This property explains why it is possible for an opera singer to shatter a glass. If they find the right note for the crystal from which the glass is made, a resonance wave is set up which can disrupt the crystalline structure and break it apart.

This property of crystals has many applications, but it has also been adopted by occultists to try to explain the role of crystals as an 'energy enhancer', as a device for focusing 'positive energy', for 'calming the mind', or even for increasing 'psychic power'. Little thought goes into these so-called explanations, and enthusiasts of the occult tend to sprinkle their descriptions with pseudo-scientific expressions they don't really understand themselves simply to impress.

In spite of what Mark Twain said in the quote at the start of this chapter, science-fiction writers may be forgiven for doing the same thing as the occultists; they are, after all, dealing in fiction, fantasy and entertainment, not trying to offer explanations for the way the universe operates. Resonance is a favourite of those who write back stories to fiction such as *Doctor Who, Star Trek* or indeed much of science fiction created during the past century. Two examples are the crystals that make the Tardis function and the dilithium crystals that power the warp drives aboard the *Enterprise* which act in some mysterious way to produce 'quantum resonance'.

One of the best fictional uses of the idea of resonance is in the story *The Shout*, written by the great novelist Robert Graves and made into a film in 1978 starring Alan Bates. In this story the lead character, Charles Crossley, claims he can kill with a shout and that he acquired this skill while living with an aboriginal tribe whose ancestors had practised The Shout since ancient times. Although the story goes into little detail about the mechanics of this mysterious talent, it is suggested that Crossley kills with a shout by applying the same sort of resonance effect as that which shatters a crystal wineglass.

The idea of using resonance has some rather prosaic applications in everyday life, as well as some very exotic ones that turn up in science fiction. Resonance lies behind the use of ultrasonics to shatter gall stones, and it has even been used to clean monuments – literally shaking the dirt into invisibly small fragments. And it is not just the fictional military designers living on alien worlds who show an interest in building resonance weapons. The idea of producing a machine that can project an invisible beam to disrupt crystal structures (or indeed any chemical unit with a regular framework) is obviously one to excite Earth-bound designers of new weapons, such as the researchers who today work for DARPA in the United States and DERA in Britain.

CONCLUSION

The Doctor is a maverick – an eccentric, rebellious Time Lord who appears to have an inexplicable fondness for humans. But, from what we know of his people, the Gallifreyans, he is certainly not typical. The Doctor likes mucking around with extremely primitive technology, but his fellow Time Lords seem to be a rather straight and serious bunch who don't really understand the way our hero thinks.

What I have referred to as Gallifreyan magic is not really that; it is extremely advanced technology. Some of the things described in this book will soon be within our reach – robots, clones, genetic manipulation, interplanetary travel. Others, though, are either truly impossible (the transdimensional property of the Tardis is exciting, but it does not sit well with the laws of physics), or they may only be achievable in the distant future. Included in this list would be the technologies behind teleportation, interstellar travel and, of course, time travel.

Some readers may consider even pondering these things as rather pointless, but then I would imagine those who feel that way would not have stuck with me this far. I'm convinced of the opposite, that there is much to be learned from considering things that lie at the border between the possible and the impossible, because, for one thing, this line is never static. What seemed impossible even a decade ago is now being reconsidered in the light of new ideas and new discoveries.

The essential ingredient in progressive thinking is imagination. This helps us to push back the barriers and to find new and exciting avenues to follow. Imagination is the most important quality any of us possess. It is, I believe, no exaggeration to say that the use of imagination is an essential requirement for the future well-being of humanity.

Imagination has served us well so far, for without it, Einstein, Newton, Darwin and all the other great pioneers of science would have failed to tease their own tiny bit of magic from the jealous grasp of Nature. And imagination will continue to serve us well in the centuries to come. It will fuel the fantasies and ambitions of future

Einsteins, Newtons and Darwins, people who may yet work out ways to acquire the technology of the Gallifreyans, to travel in time and through interstellar space, to live as long as we like, to read minds and to teleport at will. Only in this way does science fiction become science fact.

Index